Memoirs of The New York Botanical Garden

Series Editors

William R. Buck
New York Botanical Garden
Institute of Systematic Botany
Bronx, NY, USA

Lawrence M. Kelly
New York Botanical Garden
NYBG Press
Bronx, NY, USA

Launched in 1900 by The New York Botanical Garden (NYBG), and with 120 volumes published as of 2019, Memoirs has traditionally focused on taxonomy and floristics, with an emphasis on plants of the Western Hemisphere. Most titles comprised a single volume, but quite a few were multi-volume titles and constitute some of the most iconic and substantial monographic publications of NYBG. In recent years, the Memoirs series has expanded to encompass field guides and botanical history as well. All in-print volumes of Memoirs through volume 120 are available exclusively through NYBG Shop: https://nybgshop.org/nybg-press/. Volumes 121 and going forward are available exclusively through Springer.

More information about this series at http://www.springer.com/series/16366

Joel Schwartz

Robert Brown and Mungo Park

Travels and Explorations in Natural History for the Royal Society

 Springer

Joel Schwartz
Professor (emeritus) of Biology and the History of Science
College of Staten Island of the City University of New York
Swampscott, MA, USA

ISSN 0077-8931   ISSN 2662-2858   (electronic)
Memoirs of The New York Botanical Garden
ISBN 978-3-030-74861-6   ISBN 978-3-030-74859-3   (eBook)
https://doi.org/10.1007/978-3-030-74859-3

Cover images: (left) Portrait of Robert Brown by an unknown artist (detail), © The Trustees of the
Natural History Museum, London; (center) H.M.S. Investigator, commanded by Matthew Flinders—July
1802 by John Sheard Grafton (detail), 2006, from Elynor Frances Olijnyk, AHA AAMH, Taking
Possession: A Saga of the Great South Land (Marino, South Australia: Photographic Art Gallery, 2007),
©Photographic Art Gallery, 2007; (right) Portrait of Mungo Park by Atkinson Horsburgh, © Scottish
Borders Council.

This Springer imprint is published by the registered company Springer Nature Switzerland AG
The registered company address is: Gewerbestrasse 11, 6330 Cham, Switzerland

In memory of my wife Rhoda Schwartz, a dedicated science teacher, whose light will shine forever.

Preface

In the lobby of the Royal Society of London, there is a fine staircase that ascends up to the research rooms of the library. On the wooden wall lining the back of the staircase, there are the names of the presidents of the Royal Society embedded in the wood in gilt lettering, beginning with the first, mathematician Viscount William Brouncker (serving from 1662 to 1677), and ending with the current president, biophysicist, Sir Venkatraman Ramakrishnan (since 2015). The names of many illustrious scientists are found in this group, including Sir Isaac Newton (1703–1727) and Thomas Henry Huxley (1883–1885). Naturalist and botanist Sir Joseph Banks's name stands out because he had the longest tenure of any other of these figures (1778–1820).

Joseph Banks (1743–1820) was a significant naturalist-explorer in his own right. He served as Captain James Cook's naturalist on H.M.S. *Endeavour*. As a result of his contributions and social position, he became the President of the Royal Society of London, an organization of the nation's most eminent thinkers and scientists founded in the 1660s. As President of the Royal Society, Banks enjoyed an exalted status in government circles. His long service as the Society's President (1778–1820) is testimony to that fact. Thus, he enjoyed a privileged position in British science.

Banks played an important role in determining policy with respect to scientific expeditions. He helped initiate investigations in natural history growing out of voyages of exploration. He was the focus of activity in natural history and other fields of scientific inquiry in Britain during his tenure at the Royal Society. He initiated most projects, and naturalists sought his attention and approval.[1] His leadership of the Society shaped Robert Brown and Mungo Park's careers.

Banks made several important voyages on his own before his leadership of the Society. As the naturalist on Cook's first epic journey on H.M.S. *Endeavour* (1768–1771), he helped prepare several young naturalists for botanical exploration, particularly his assistant, Daniel Solander (1733–1782), who originally was a

---

[1] See Patrick O'Brian's *Joseph Banks. A Life* (Boston: David R. Godine, 1993) and Susan Faye Cannon's *Science in culture: the early Victorian period* (New York: Science History Publications, 1978).

disciple of Linnaeus but chose to move to Britain and work under Banks. Many other naturalists were motivated to follow in Banks's footsteps. The varied collection of plants gathered from colonial territories influenced the development of botany. It has additional relevance for Americans because the origins of botany in the United States can, in part, be attributed to the work of Banks and his disciples.

Banks's experience as the naturalist on Captain Cook's first voyage around the world made him an advocate for journeys of exploration. He helped shape the development of botany, earth science, art, horticulture, and other aspects of British science and culture in the late eighteenth and early nineteenth centuries after his voyage on the *Endeavour*. All efforts to conduct travels on behalf of natural history in Britain went through him, and he had a role not only in the implementation of such projects but also in gathering the results of such journeys: the flora, fauna, fossils and other geological samples, and seeds of many of the plants that were collected. His greatest legacy was the effect he had on the naturalists who followed him into a career of exploration in natural history, particularly Robert Brown and Mungo Park.

Banks advanced international cooperation in science by maintaining contact with France during the French Revolution and the Napoleonic Wars. In spite of this open-minded attitude, the tensions from the Napoleonic Wars as well as other international events that occurred during most of his tenure as President often tested his belief that there should be no borders as far as science was concerned. This was borne out when the man he selected to be Captain of the *Investigator*, Matthew Flinders, was held captive by the French when hostilities between the British and French resumed before Flinders could safely return to England. Although Banks was unable to provide much assistance to Flinders, he never wavered from his belief that science should be a collaborative effort among scientists from many nations. Nevertheless, his role in organizing expeditions resulted in assisting Britain's interests in furthering its imperial ambitions. Sir Joseph Banks, therefore, was a central figure in the great expansion of scientific exploration in the late eighteenth and early nineteenth centuries and played a central role in the events discussed in this book.

Kew Gardens, NY, USA   Joel Schwartz
Swampscott, MA, USA

Acknowledgments

After my retirement from full-time teaching at the City University of New York, I conducted research for *Robert Brown and Mungo Park: Travels and Explorations in Natural History for the Royal Society*. At about the same time, I was also completing my previous book, *Darwin's Disciple, George John Romanes, A Life in Letters*. During the past decade, I have been busily working on *Travels and Explorations*, describing the lives and careers of Scottish naturalists Robert Brown and Mungo Park. I have been the beneficiary of support and guidance during this time, and therefore I am indebted to those who gave me their advice and assistance.

The New York Botanical Garden and its excellent LuEsther T. Mertz Library have played an important role in shaping my appreciation of the botanical sciences and have been very supportive in allowing me to complete this book. Numerous people at the Mertz Library went out of their way to provide me with assistance, most particularly, Susan Fraser, the Thomas Hubbard Vice President and Director of the Library. She was instrumental in helping me publish this work. William R. Buck, Senior Curator Emeritus of the Institute of Systematic Botany of the New York Botanical Garden and Editor, Memoirs, has provided me with his expertise and insight. Michael Brown of the NYBG Press Office helped facilitate the publication of this work and made many useful suggestions. Others there, such as Reference/Circulation Librarian Marie Long, were also very helpful.

My parents were a source of inspiration in guiding me toward serious pursuits. My father, George Schwartz, was a high school teacher, a college biology instructor, a naturalist, and a skilled microscopist. My interest in Robert Brown may have evolved from the range of my dad's interests. My mother, Hannah Schwartz, was dogged in her efforts in getting me to develop serious interests, taking my sister and me on long subway trips to museums, to pottery classes, and to music lessons. Many times, she urged me to read such works as Zsolt Harsanyi's *The Star Gazer*, understanding that books like this classic would broaden my perspective.

My teachers at Stuyvesant High School were also important in my early education. The New York City school system was instrumental in guiding young people to make the most of their abilities. Institutions such as the public schools in America are under assault today, and the degradation of this valuable resource will compromise the future of our country. Several professors at the University of Rochester helped shape my career. The late Professor Hayden White in the history department helped me to develop a deep appreciation of intellectual history. Professor Richard Lewontin, the renowned geneticist, furthered my development in the biological sciences.

In Britain, Gina Douglas, the former librarian and archivist at the Linnean Society, and Lynda Brooks, the current head librarian, have been responsive to my questions. I had the assistance of numerous people at the British Library, the Natural History Museum and Botany Department, the Royal Society, and the Royal Botanic Gardens, Kew. I am indebted to the New York Public Library Rare Book Room; the Linnean Society of London; the Museum of Natural History in London; the Scottish Borders Council; the Jens Nachtigal website; the Hydrographic Office of the United Kingdom, London; the Ministry of Defense Collection, London; the National Picture Library of Australia; the Western Australian Herbarium (Department of Parks and Wildlife); the Australian Botanic Garden; the University of Sydney Archives; and the State Libraries of Victoria, New South Wales, and South Australia for the use of their illustrations and maps.

For much of this time, I have had Wertheim Study privileges at the New York Public Library and with Melanie Locay, Research Study Liaison, and Rebecca Federman, Research Services Coordinator, who continue to provide support in allowing me to conduct my research and writing in relative peace and quiet. For the past few years, I have been part of a writing group at the offices of the Professional Staff Congress of CUNY, composed of people from different fields. Special mention should go to Lecturer Constance H. Gemson, the tireless leader of the group, and Professors Irvin Sam Schonfeld, David Kotelchuck, and Helen Yalof, whose useful comments have been quite helpful. Our monthly meetings have brought me a great deal of joy as we critique each other's writing projects.

Most of all, special thanks to my late wife Rhoda, whose support has been very valuable in writing this book, and my sons Mark and George who have given me a great deal of advice and help while they tend to their own families and careers.

Contents

Introduction: Collecting, Observing, and Describing the Natural World

Prior to publication of Charles Darwin's *Origin of Species* (1859), the predominant view was that living things were the products of God's creation and had not appreciably changed since the biblical account related in Genesis. While there were naturalists and philosophers since antiquity that questioned the prevailing notion that species were immutable, these ideas were not able to take root until the increase of governmentally funded geographic expeditions in the late eighteenth and early nineteenth centuries. The narrow experience of most naturalists before the mid-eighteenth century explains, in part, why belief in the immutability of species prevailed.

Travel to exotic places increased in intensity. Voyages of exploration allowed naturalists to travel to exotic lands, ushering in an "age of discovery." The explorer-naturalists on board ships encountered living things not seen before. The natural specimens they brought back, and the discovery of fossil forms of extinct organisms, weakened the arguments of those who maintained their belief in the literal interpretation of the Bible. Such voyages increased in the middle of the eighteenth century with the development of the marine chronometer, making it possible for ships at sea to determine the longitude of their location. Naturalists who planned to embark on voyages no longer had the same concerns as their predecessors because they were able to determine their exact position at sea. Alexander von Humboldt, Charles Darwin, Alfred Russel Wallace, and others had the opportunity to explore exotic, often unfamiliar, territory, leaving the cultivated lands they grew up in. They observed the ferocity in nature with abundant varieties of many unfamiliar forms of life and geological phenomena in their original setting, making it more difficult to sustain the view that living forms were immutable.

Beginning in the seventeenth century, voyages of discovery resulted in exploration of much of the previously unknown world. Early pioneers included German-born naturalist and illustrator Maria Sibylla Merian (1647–1717), who studied plant life in Dutch Guiana (Suriname), and the British physician and naturalist Hans Sloane (1660–1753), who traveled to Jamaica and collected 800 species of plants and animals, forming the bulwark of the British Museum's collections.[1] The German-born physician and botanist Paul Hermann (1646–1695) studied native plants of Ceylon (Sri Lanka) for the Dutch East India Company. Later on, the Dutch Governor of Ceylon, Johan Gideon Loten (1710–1789), also collected flora, fauna, and minerals for the Dutch East India Company.

Captain James Cook's voyages around the world were partly the result of these earlier adventures. These voyages spurred many advances in the natural sciences, ethnology, geography, and astronomy.[2] Sir Joseph Banks (1743–1820), Cook's naturalist on H.M.S. *Endeavour*, became the President of the Royal Society of London, an organization of the nation's most eminent thinkers and scientists established in the 1660s. Under Banks's leadership, the Society played an important role in shaping the careers of Robert Brown and Mungo Park. Georg Forster (1754–1794), a naturalist, illustrator, and humanist, succeeded Banks on Cook's later voyages. Forster promoted enlightenment thought and attempted to introduce liberal ideas by participating in the development of the short-lived democratic Mainz (German-speaking) Republic. Ultimately, he fled, spending the later part of his life in exile in Paris. The adventures of these seventeenth- and eighteenth-century naturalists inspired naturalist-explorers such as Brown and Park.

Robert Brown (1773–1858) and Mungo Park (1771–1806) were two explorer-naturalists who played a pivotal role in the development of natural history. Their adventures and investigations in natural history created a fertile environment for breakthroughs in taxonomy, cytology, and eventually evolution. Brown's pioneering work in plant taxonomy enabled biologists to look at the animal and plant kingdoms differently, and Park's journeys led to significant discoveries of a previously unknown world. Brown's and Park's adventures formed a bridge to such expeditions as Charles Darwin's voyage on H.M.S. *Beagle* (1831–1836), which led to a revolution in biology and the full explication of the theory of evolution. Thomas Henry Huxley's lesser known trip as naturalist on H.M.S. *Rattlesnake* (1846–1850) allowed him to revise the taxonomy of animals, particularly invertebrates.

Brown and Park came from similar backgrounds. They were born and grew up in the same region of southern Scotland and were strongly influenced by the ideas emanating from the Scottish Enlightenment of the late eighteenth century. This period was marked by rich intellectual and scientific development, growing out of scientific societies and Edinburgh University. As young men, they both explored the rugged hills in southern Scotland as well as its Highlands in search of native plants. They both went to Edinburgh University, the center of Enlightenment ideas, for

---

[1] James Delbourgo discusses the impact of Hans Sloane's acquisitions in *Collecting the World. The Life and Curiosity of Hans Sloane* (London: Penguin Books, 2018). [First published by Harvard University Press as *Collecting the World. Hans Sloane and the Origins of the British Museum* in 2017.]

[2] James Cook. *The Journals of Captain Cook, Prepared from the Original Manuscripts by J.C. Beaglehole for the Hakluyt Society, 1955–67* (London: Penguin Classics, 1999). This work, selected and edited by Philip Edwards, provides a firsthand account of Cook's adventures, with both the background and impact of these journeys.

their medical studies at roughly the same time. However, Park and Brown were quite different in temperament, and they approached scientific investigations in a different manner. Park, a handsome man with considerable charm, was not reluctant to speak his mind. Brown was rather diffident and often reluctant in advancing his ideas but was a meticulous worker. Although they shared similar experiences and associations in Edinburgh, they never referred to one another in their correspondence with others. Their lives and careers demonstrate how scientists from very similar backgrounds can develop independently from one another.

Park was Banks's original choice to be the naturalist for the expedition to Australia, but he refused the position after the Admiralty, also a sponsor of these voyages, did not offer him a proper salary for his services. This refusal proved to be a fortuitous circumstance for Brown, his career, and indeed the course of natural history, because Brown's dedication demonstrated that he was the right man to conduct the exceedingly difficult work demanded of him. But initially that was not clear to Banks. At the time, Park seemed to have greater promise than Brown. Park's early career was quite heroic, and his escape from captivity during his first trip to Africa added to his legend. Park, perhaps, reminded Banks of himself as a younger man, another factor in his appointment to carry out investigations in Africa.

While Park went on his first African journey, serving more as an explorer than naturalist, Brown was busily engaged in field studies at the time in his native Scotland (as well as in other parts of the British Isles), observing and describing many varieties of plants he encountered. Soon after Brown's return, Park went to Africa again and made critical observations in ethnology and geography. Park's last expedition to Africa stands in contrast with Brown's experience in Australia. His death in Africa occurred while he was exploring the Niger River region, an area previously unknown to Western societies.

Brown's service as naturalist on H.M.S. *Investigator* was critical in Brown's development as a scientist. Commanded by Matthew Flinders, the *Investigator* surveyed the southern and northern coasts of Australia (called New Holland at the time, 1801), i.e., circumnavigation of the entire continent.[3] Ably assisting Brown on the *Investigator* was the Austrian-born botanical and zoological draftsman, Ferdinand Lucas Bauer (1760–1826), who was very much like his brother, Francis (Franz) Andreas Bauer (1758–1840), a uniquely skilled scientific illustrator and observer of natural history.[4]

---

[3] Flinders is recognized as the inventor of a device used to counteract the magnetism of the ironclad ships, the "Flinders bar." While earlier navigators had observed errors in their compasses, which could not be explained by magnetic variation, Matthew Flinders was the first to make a systematic investigation of the problems caused by the presence of iron in the ship. He suggested that it would be possible to correct such errors by using a vertical unmagnetized iron bar near the compass to counteract deviation caused by magnetic induction in the soft iron of the ship. Although such devices were not used until the second half of the nineteenth century, they were called Flinders bars in recognition of the originator of the idea, although Flinders had not actually used such a device.

[4] William T. Stearn, "Franz and Ferdinand Bauer, masters of botanical illustration," *Endeavour* 19 (1960): 27–35.

The period between Brown's return to England and Banks's death and immediately afterwards (from 1805 to 1826) was perhaps Brown's most productive period, highlighted by his published works. Initially, this work was conducted with Banks's support, although Brown worked increasingly on his own. Brown published the majority of his work during this period of time. Brown also facilitated the transfer of Banks's collections, library, and various herbaria (collections of preserved plant specimens) to the British Museum, allowing him to gain the position of curator of these materials and enabling him to receive a reliable source of income.[5] Brown's discoveries of Brownian movement, the phenomenon of protoplasmic streaming, and the nucleus, the dark-staining body present in cells, are perhaps the most notable of his achievements and highlight his skill in microscopy.

Brown's later travels and fieldwork, particularly his work in taxonomy, led him to adopt a more natural system of classification instead of the artificial one of Linnaeus. Brown's position as naturalist on H.M.S. *Investigator*, very much like Thomas Henry Huxley's work on the classification of animals later on, led him to revise plant taxonomy. The evidence Brown amassed from his Australian travels supported his revision of plant classification. The collections that Brown received (after his return from Australia), from the East Indies, Africa, and Central and South America from other naturalists, reinforced the changes he made in plant taxonomy (classification) and demonstrate his contributions in the more theoretical aspects of biology.

Brown had a positive influence on other naturalists such as Charles Darwin and Joseph Dalton Hooker (1817–1911). Before Darwin sailed on H.M.S. *Beagle*, Brown showed him how to use the microscope in his investigations. Years after Charles Darwin returned to England from his epochal voyage on H.M.S. *Beagle*, he recalled, "I saw a good deal of Robert Brown, 'facile Princips Botanicorum,' as he was called by [Alexander von] Humboldt. He was remarkable for the minuteness of his observations, and their perfect accuracy. His knowledge was extraordinarily great, and much died with him, owing to his excessive fear of ever making a mistake." Darwin continued, "I called on him, two or three times before the voyage of the *Beagle*, and on one occasion he asked me to look through a microscope and describe what I saw. This I did, and believe now that it was the marvellous currents of protoplasm in some vegetable cell."[6]

Darwin's correspondence, particularly with his close associate, Joseph Dalton Hooker, cited Brown numerous times. It illustrates the regard Darwin and his associates had for Brown's abilities as a naturalist in addition to capturing the essence of Brown's single-mindedness in pursuing his scientific interests. Despite not being particularly close—Darwin was a good deal younger than Brown, and there is no

---

[5] Brown agreed to the transfer of the Banks herbarium and library to the British Museum and was officially appointed Under-Librarian with the designation of Keeper of the Sir Joseph Banks Botanical Collection, i.e., the Banksian Collection.

[6] Charles Darwin, *Autobiography of Charles Darwin, and Selected Letters*, ed. by Francis Darwin (New York: Dover Publications, 1958), p. 36. Darwin wrote these remarks in 1875, and they are in a collection of his reminiscences.

record of correspondence between them—Darwin's description of Brown is quite appropriate. Naturalists held Brown in great esteem. He enjoyed a fine reputation among naturalists inside and outside of Britain. Botanists with the opportunity to travel on voyages of exploration, microscopists and cytologists, naturalists engaged in plant geography, and taxonomists, all admired his work.

Mungo Park died 3 years before Darwin's birth. Thus, his life had little impact on Darwin and other contemporary naturalists save those whose imagination was captured with narratives of Park's adventures and heroism. His premature death limited his accomplishments in the sciences.

Brown helped Hooker obtain his position as naturalist on H.M.S. *Erebus* and *Terror*. Hooker served on *Erebus* and *Terror*, ships that explored the southernmost regions of the world in the 1840s. His opportunity to take part on this voyage was the result of Brown's friendship with Hooker's father, William Jackson Hooker (1785-1865), the first director of the newly reincorporated Kew Gardens.[7] Brown's relationships with contemporary naturalists further elucidate his importance in the development of nineteenth-century natural history.

Therefore, Robert Brown is an important but often neglected figure in science. In spite of his nineteenth-century contributions to natural science and the relevance of his work to modern natural history, he has been overlooked. His work in natural history, taxonomy, and cytology, highlighted by his discovery of the phenomenon of random movement of particles and protoplasmic streaming, and his role in the discovery of the cell nucleus, brought him respect and admiration from peers. Yet he remains largely unknown today, except for the discovery of the random motion of particles in a suspension ("Brownian movement," named in his honor). Through skillful use of the microscope, he examined the fine structure of plants, critical in allowing him to support a natural system of classification. However, his retiring nature, even with associates—he was naturally shy and rarely talked about himself—hampered him from receiving greater recognition despite his prowess as an acute observer of natural phenomena.

Although Mungo Park's career in the sciences did not nearly match Brown's, Park's role as an explorer and adventurer left an indelible mark on British history. Initially, Park's scientific career appeared to be more promising than Brown's, and he had an indirect but vital role in Brown's career. Their early lives growing up in rural Scotland and their exposure to the intellectual ferment in late eighteenth-century Edinburgh played a critical role in directing their careers in the natural sciences.

---

[7] After Banks's death, the Royal Gardens at Kew had sunk to the status of a royal pleasure garden. The Parliament decided in 1840 that Kew Gardens be developed into a national botanic garden, and it became a public institution under the Board of Woods and Forests. It was decided that a director must be appointed, who would be capable of carrying out the intention of the Parliament. Thus, in March of 1841, Sir William Hooker was appointed to this position. See Raymond G. C. Desmond's, "The Hookers and the Development of the Royal Botanic Gardens, Kew," *Biological Journal of the Linnean Society* 7 (1975): 173–182.

Chapter 1
Scientific Ferment in Late Eighteenth-Century Edinburgh

Contents

Edinburgh and the Scottish Enlightenment

Brown and Park benefited early in their careers from the fertile intellectual climate generated in late eighteenth century and early nineteenth-century Scotland, particularly Edinburgh, during the height of the "Scottish Enlightenment." Brown and Park's immersion in the atmosphere of late eighteenth-century Edinburgh was fortuitous in shaping their early development. It was a period of great intellectual activity, with remarkable advances in science and technology—the growth and development in the geological sciences was one such example—as well as in literature and the arts.

During this period, attention to scientific ideas in Scotland flowed from the activity of the Edinburgh University Medical School. Under the guidance of the school's professors, and with the support of private physicians, advances in medicine created interest in a wide range of fields, including chemistry, physiology, and botany. As interest in the large range of scientific fields grew, scientific activity became more professionalized.[1] The favorable climate also assisted the development of medical societies such as the Society for the Improvement of Medical Knowledge (founded

---

[1] The Society for the Improvement of Medical Knowledge consisted of "the medical professors of the University and many of the leading members of Edinburgh Colleges of Physicians and Surgeons .... The professionalization of Edinburgh medical science is further indicated by the proliferation of specifically medical societies and clubs ... the convivial Aesculapian Club (1773), the Harveian Society (1782) ...." Steven Shapin, "The Audience for Science in Eighteenth Century Edinburgh," *History of Science*, 12 (1974), 95–121, 98. Robert Chambers was another notable figure nurtured by this atmosphere. See Joel Schwartz, "Robert Chambers and Thomas Henry Huxley, Science Correspondents: The Popularization and Dissemination of Nineteenth Century Natural Science," *Journal of the History of Biology* 32 (1999): 343–383.

© The Author(s), under exclusive license to Springer Nature Switzerland AG 2021
J. Schwartz. *Robert Brown and Mungo Park*, Memoirs of The New York Botanical Garden 122. https://doi.org/10.1007/978-3-030-74859-3_1

in 1731), the Aesculapian Club (begun in 1773), and Harveian Society (established in 1782), among the more prominent groups.[2]

Commercial activity also aided Edinburgh's growth as a center for medical education and research. Local trade and mercantile guilds understood that by supporting scientific investigation, promising students would not be so easily drawn away by London's energy and the prosperous commercial and technological center of nearby Manchester.[3] Scientists were influenced by the stimulating public life in all of Great Britain, especially in those regions most directly affected by the industrial revolution. Their activity was inextricably linked with the success of commerce and the new wealth it generated. Business interests assisted scientific inquiry; they understood that scientific innovation had a favorable impact on economic development.[4]

The Royal Society of Edinburgh was founded in 1783. Like the medical societies, it acted as a spur for scientific development and was also a reflection of the growth in Scottish science. The founding of the Society marked a change in Scotland, with the natural sciences becoming more specialized and medicine more professionalized. The study of botany grew in importance economically and became an independent branch of natural history. Because the Royal Society was founded to foster all scientific disciplines, it was an environment conducive for Brown to be able to pursue his interest in natural history, particularly botany.[5]

---

[2] Jacqueline Jenkinson indicated that the "development of the medical profession in Scotland is reflected in the rise of medical societies," *Scottish Medical Societies, 1731–1939, Their History and Records* (Edinburgh: Edinburgh University Press, 1993), p. 2. She has traced the development of these groups from their establishment in the eighteenth century to modern times.

[3] Shapin cited the role of Edinburgh's trade and merchant guilds in fostering the growth of Scottish medical teaching and clinical research, "The Audience for Science in Eighteenth Century Edinburgh," p. 97. Jack Morrell focused on Glasgow's expansion as a scientific center and contrasted it with Edinburgh's, finding that it was based on "the entrepreneurial skill of her businessmen … merchants and manufacturers," while Edinburgh's growth represented by the Royal Society of Edinburgh benefited more from "the contributions made by the University professoriate, the medicals, the lawyers, and the lowland landed gentry," Jack Morrell, "Reflections in the History of Scottish Science," *History of Science* 12 (1974): 81–94, p. 89. Republished in *Science, Culture and Politics in Britain, 1750–1870* (Aldershot, Hampshire: Variorum, 1997), pp. 81–94.

[4] Agriculture also fostered the development of areas of science that were medically related; one such example was the development of the smallpox vaccine. Chemistry and geological science also flourished in this atmosphere.

[5] Paul Wood discusses the development of the sciences during the Scottish Enlightenment, and the role of the Royal Society of Edinburgh in "Sciences in the Scottish Enlightenment," *The Cambridge Companion to the Scottish Enlightenment*, Alexander Broadie, ed. (Cambridge: Cambridge University Press, 2003), pp. 107–110. Scotland was also affected by the marked increase of exploration that took place in the eighteenth century as it became more integrated with the rest of Britain. The search for the natural resources of distant lands included plants and plant products. Plants were studied and cultivated to see if they had economic or medicinal value. The study of botany was enhanced by herbarium specimens and botanical illustrations, and advances in taxonomy that simplified the description of plants. As a consequence, botany developed into a separate branch of natural history.

Edinburgh was exhilarating for young men like Brown and Park. The intellectual climate had as much of an impact on Brown as did his formal education. Edinburgh from the middle to the late eighteenth century was a place marked by exciting new ideas in the arts, literature, and particularly the sciences, i.e., the "Scottish Enlightenment," a period from approximately the 1740s to the early 1800s. The movement's origins can be traced to when Scotland joined England to become part of Great Britain (the Act of Union in 1707). The Highland clearances that began in the early part of the eighteenth century—whether they were a direct consequence of the Act of Union or a manifestation of changes in Scottish society is not certain— involved the forcible removal of people who practiced a more diversified agriculture to make room for sheep farming. This depopulation of northern Scotland destroyed clan society and was accompanied by a great loss of life, resulting in the migration of many Scots to North America. Despite the hardships that followed the 1707 Act of Union (e.g., the Highland Clearances and the depopulation caused by the Clearances), some benefits accrued to the Scots. Although Scotland lost some of its autonomy, a positive aspect of this annexation was that Scotland opened up to fresh ideas.[6] Scottish thinkers began questioning assumptions previously taken for granted, and developed a uniquely practical branch of humanism.

An early important figure in this movement was philosopher Francis Hutcheson (1694–1746) who was at Glasgow University from 1729 to 1746. Hutcheson's leading idea was that virtue is what brings the greatest good to the most people.[7] Many outstanding figures emerged during this period of Scottish history, e.g., Scotland's beloved writers, poet Robert Burns and novelist Sir Walter Scott, architect Robert Adam, philosopher and skeptic David Hume, and economist-philosopher Adam Smith, to cite just a few examples.

The Scottish Enlightenment eventually shifted its focus from intellectual and economic matters to scientific concerns, highlighted by the contributions of James Anderson, a farmer and lawyer with a strong interest in agronomy who applied scientific principles to farming. The "Scottish Enlightenment" accounted for a disproportionately large number of contributions to British science and letters, including such notable individuals as the founder of modern geology James Hutton, engineer and inventor James Watt, chemist Joseph Black, naturalist John Walker, and later on physicist James Clerk Maxwell.[8]

---

[6] Scotland reaped the economic benefits of free trade within the growing British Empire. It also benefited in having established Europe's first public education system since classical times. There is a question about how much the "Act of Union" led to the "Highland Clearances." The Scottish legal system allowed landlords to exploit the system. Scottish farmers may have been better served by the English legal system.

[7] Philosopher David Hume was concerned with the nature of knowledge, and embraced the scientific method with ideas about evidence, experience, and causation, in addition to modern attitudes concerning the relationship between science and religion.

[8] Described in more detail in Alexander Broadie's *The Scottish Enlightenment: The Historical Age of the Historical Nation* (Edinburgh: Birlinn Ltd., 2001).

Brown's immersion in the atmosphere of late eighteenth-century Edinburgh influenced his development as a scientist of the first rank. Edinburgh University Medical School became the center of scientific ideas in Scotland. With guidance from the school's professors and support of private physicians, developments in medicine created interest in a wide range of fields, including chemistry, physiology, and botany. The development of medical societies previously mentioned (e.g., the Society for the Improvement of Medical Knowledge, the Aesculapian Club, and Harvian Society) was also a sign of professionalization in the medical sciences in Edinburgh during this period.[9] Because Brown was not born into privilege and needed to support himself, he welcomed the opportunities created by these developments in the sciences. Brown was drawn to making a career in the sciences not only out of necessity but also because he was fascinated by the growth of many branches of the sciences in late eighteenth-century Edinburgh.[10] Scientific inquiry was no longer restricted to rich amateurs but was becoming a legitimate means of supporting those who actively engaged in scientific investigation.

The symbiotic relationship between science and commercial interests enhanced Edinburgh's growth as a center for medical education and research. Business interests helped promote scientific activity because they understood that scientific innovation could impact economic development favorably. Agriculture fostered the development of science that was medically related. One such example was the development of the smallpox vaccine. Chemistry and geological science flourished in this atmosphere.[11] The founding of the Royal Society of Edinburgh in 1783, shortly before Robert Brown's arrival there less than a decade later, was a favorable development for young men like Brown. Because the Society's commitment was to support all scientific disciplines, the study of botany also reaped the benefits of this favorable climate.[12] Brown felt free to pursue his interest in natural history, particularly botany, and he did so with great enthusiasm.

---

[9] Steven Shapin, "The Audience for Science in Eighteenth Century Edinburgh," *History of Science* 12 (1974): 95–121.

[10] Jacqueline Jenkinson, *Scottish Medical Societies, 1731–1939, Their History and Records* (Edinburgh: Edinburgh University Press, 1993), p. 2. See note 2 on page 244.

[11] Shapin, "The Audience for Science in Eighteenth Century Edinburgh," p. 97. Morrell, "Reflections in the History of Scottish Science," *History of Science* 12 (1974): 81–94, p. 89. Republished in *Science, Culture and Politics in Britain, 1750–1870* (Aldershot, Hampshire: Variorum, 1997), pp. 81–94. See note 9 on pages 244–245.

[12] Fredrik Albritton Jonsson expresses a somewhat different view in *Enlightenment's Frontier, The Scottish Highlands and the Origins of Environmentalism*, suggesting that the Scottish Enlightenment had its origins in the mountains and moss bogs of the Scottish Highlands. He maintains that Environmentalism began in the Highlands, which according to Jonsson helped spawn the Enlightenment, more than the Edinburgh University and the Scientific Societies (New Haven & London: Yale University Press, 2013).

Chapter 2
Scientific Exploration During Voyages of Discovery

Contents

The age of exploration developed from scientific discoveries that took place in England's two great universities, Cambridge and Oxford. In the sixteenth and seventeenth centuries, the study of geography helped generate the growth of new ideas. Mathematical geography flourished at Corpus Christi and St. John's College, Oxford, and Peterhouse and Corpus Christi College, Cambridge. The work of England's foremost mathematical geographer, Edward Wright (1561–1615), and the contributions of mathematician, astronomer, and physicist, Thomas Harriot (1560–1621), aided nautical exploration by putting the theoretical information they learned to practical use in navigation.

In 1589 Wright received permission to take a leave from Cambridge to accompany the Earl of Cumberland on an expedition to the Azores. He already enjoyed considerable standing in the field of mathematical navigation, and his critique of the sea charts of the time—which he called "an inextricable labyrinth of error"—helped cement his reputation. When Wright left Cambridge, he became a mathematical lecturer; the need for lecturers in navigation was acute at the time. In 1614, while serving as a lecturer, the East India Company took over patronage, paying Wright an annual salary of £50. Wright was also a surveyor for the New River project whose purpose was to bring water to London. Wright's *Certain Errors* was justification for the Mercator-map projection, up to that time, the most important advance in marine cartography.

After completing his undergraduate studies at Oxford, Harriot was asked by Sir Walter Raleigh—who needed an expert in cartography and the theory of oceanic navigation—to accompany him on a colonizing expedition to Virginia in 1585.

© The Author(s), under exclusive license to Springer Nature Switzerland AG 2021
J. Schwartz, *Robert Brown and Mungo Park*, Memoirs of The New York
Botanical Garden 122, https://doi.org/10.1007/978-3-030-74859-3_2

Harriot investigated the life, language, and customs of the inhabitants of Virginia while he surveyed the coast, islands, and rivers.[1]

The advances in mathematical geography and cartography fed the scientific revolution and spurred naturalists to take part in voyages of discovery, allowing them to collect many species of flora and fauna hitherto unknown in the developed world. Examples of voyages that resulted in discoveries in natural history were Sir Hans Sloane's travels to Jamaica (1687–1689), notable for the development of the chocolate trade; Paul Hermann's journey to Ceylon (1672–1675); later on Governor Johan Gideon Loten's work in Ceylon (1751–1757) on behalf of the Dutch East India Company; Maria Sibylla Merian's explorations in Suriname (1699–1701); and American-born William Bartram's expeditions in North America (1753–77).[2]

Carl von Linné (Carolus Linnaeus) dominated eighteenth-century natural history because of the popularity of his system of classifying animals, plants, and minerals, i.e., the Linnean system based on the sexuality in plants as well as his implementation of binomial nomenclature, giving scientific names to organisms. Like many naturalists, he studied medicine, and eventually traveled to Lapland. Linnaeus made no dramatic discoveries. His scholarship was considered below that of a scientist of the first rank, and he was a vain man as well as a remarkable self-promoter. Yet his 1729 paper, "Praeludia Sonsaliorum Plantarum," describing the sexuality of plants—quite controversial at the time—and then his travels to Lapland by horseback, helped solidify his reputation, together with his *Flora Lapponica*, the first work in which he employed his system of classification. Linnaeus's passion was in finding and cataloging the wonders of nature. In the view of Linnaeus and other naturalists of the time, this was a pious activity, one that glorified God who made nature just as it was and would keep it that way as long as it pleased Him.[3] Linnaeus became celebrated all over Europe, and the plants he collected are found in herbaria in the major cities—including London, where these collections remain preserved in the vaults of the Linnean Society.

The travels of Sloane, Hermann, Meriam, and Loten and insight of Linnaeus did not lead them to reject the standard religious view that the world and the plants and animals populating it were the immutable creations of God. They did view this activity—with their detailed observation, recording, and interpretation of natural phenomenon—as a legitimate and worthy pursuit. The voyages of Louis-Antoine de Bougainville, who had a botanist on board his voyage in 1766–1769, and Napoleon's

---

[1] Most of Harriot's original work was lost, but some of it was reproduced in Richard Hakluyt's *Principal Navigations* 3 vols. (London 1598–1600). This work incorporates an account of the Virginia expedition, *A Brief and True Report of the New Found Land of Virginia* (London, 1588).

[2] Tony Rice discusses these explorations in his *Voyages of Discovery: Three Centuries of Natural History Exploration* (New York: Clarkson Potter, 1999 [For the Natural History Museum of London]. Also see Diane Welebit's note, "The 'Wondrous Transformations' of Maria Sibylla Merian," for *Garden* [Publication of the New York Botanical Gardens], March/April 1988, 11–13.

[3] Tony Rice. *Voyages of Discovery*, p. 14. Also a brief but useful discussion of this aspect of Linnaeus's character is in Bil Gilbert's "The Obscure Fame of Carl Linnaeus," *Audubon* 86 (September 1984), 102–114.

sponsored expedition that sent Nicolas Baudin to survey the Australian coast in 1800 with 24 scientists in tow—all of whom Napoleon had a hand in selecting—are further examples of how support of sympathetic governments and private organizations assisted such exploration and enhanced science not only for the good of the state but also for economic gain. This caught the attention of the British Government, and they urged Joseph Banks to organize an expedition to survey the Australian coasts because of his central position in British science.

Banks's Role in Advancing Exploration in Natural History

Joseph Banks stood out among naturalist-voyagers. He made several important voyages on his own before his presence as naturalist on Cooks's first epic journey; his travels on H.M.S. *Endeavour* (1768–1771) helped prepare several young naturalists in botanical exploration, particularly his assistant, Daniel Solander (1733–1782), who previously enjoyed the patronage of Linnaeus but decided to throw his lot in with the British and Banks. Many other naturalists were motivated to follow in Banks's footsteps; the varied collection of plants gathered from colonial territories influenced the development of botany, and has additional relevance for Americans because the origins of botany in the United States can, in part, be attributed to the work of Banks and his disciples. Like naturalists earlier in the century, Banks understood the role "economic ventures" had in initiating botanical exploration.[4]

---

[4] Londa L. Schiebinger indicates that "Sir Hans Sloane at one end of the eighteenth century [chocolate trade] and Sir Joseph Banks at the other joined economic ventures to botanical exploration." She describes what she calls "applied botany" as "an essential part of the projection of military might into the resource-rich East and West Indies," crucial in "colonizing efforts in tropical climates," *Plants and Empire. Colonial Bioprospecting in the Atlantic World* (Cambridge Mass. and London, England: Harvard University Press, 2004), p. 5. Toby and Will Musgrave indicate that Banks understood the economic importance of collecting and studying "the flora of Britain's colonial heritage," but it "was mixed with a desire to preserve the ecosystems in which they flourished," *An Empire of Plants: People and Plants that Changed the World* (London: Cassell & Co., 2000). Banks was particularly concerned about the wholesale "felling of the cinchona trees for their valuable bark," p. 7. Nigel Rigby reviews how Banks as a young naturalist on the *Endeavour* was not so much "fascinated" by the "domestication of the wilderness, nor its benefits to the Tahitians, nor even the extension of England's imperium, but whether the planting would succeed." Later on, "as the effective director of Kew Gardens Botanic Gardens, Banks would later become deeply involved in the global transportation of live plants and seed, and a powerful supporter of using botanical science within an imperial context, but at this point in his career he is posing purely practical questions: whether the seeds have survived the ocean voyage." It could be argued that Brown was very much like the young Banks and would remain so, particularly later on when he was deeply concerned about the safe transport of what he collected back to Britain. "The Politics and Pragmatics of Seaborne Plant Transportation, 1769–1805," pp. 81–100, in *Science and Exploration in the Pacific European Voyages to the Southern Oceans in the Eighteenth Century* Margarette Lincoln, ed. (Woodbridge, Suffolk and Rochester, NY: The Boydell Press in Association with the National Maritime Museum, 1998), p. 84.

Banks's initial training as a botanist was enhanced by his friendship with Philip Miller, custodian of the Chelsea Physic Garden. Banks had the luxury of wealth and social status that allowed him to indulge his interests. His privileged background, his previous experience on H.M.S. *Niger*, which sailed to Iceland, Newfoundland, and Labrador (1766–1767)—where he vigorously collected many plants and animal specimens—and his own association with Linnaeus were responsible for his position as naturalist on H.M.S. *Endeavour*. Numerous botanical and zoological specimens were brought back to England from previously uncharted lands. John Ellis, a merchant and amateur naturalist who assisted British botany by importing seeds from many lands, wrote to his friend Linnaeus while Banks and Solander were preparing for their voyage on the *Endeavour*: "No people ever went to sea better fitted out for the purpose of Natural History."[5] Banks's wealth was also a benefit in pursuing such voyages of exploration; he subsidized these journeys out of his own pocket.

Banks's experience on the *Endeavour*, particularly the botanical investigations and the information spawned by the naturalists, artists, draftsmen, and collectors serving under him, the herbarium and library he assembled upon his return, and his relationship with Kew Gardens ("the King's Garden"), the first botanic garden in Europe, were some of the gains realized by his voyages. After his return from his travels, the Royal Botanic Garden, Kew, became an important botanical research center. Banks was seen as the "unofficial founding father of Kew Gardens."[6]

Banks's assistant, Daniel Solander, as well as such disciples as Jonas Dryander and Robert Brown are especially noteworthy examples of the positive gains resulting directly from Banks's involvement. Several collectors, men from more humble backgrounds such as David Nelson, a gardener at Kew, became a collector and assistant naturalist on Cook's third voyage (1776–1780) on H.M.S. *Resolution* and *Discovery*.

Due to the premature death of his most trusted disciple, Solander, Banks never pulled together his research and the vast material collected to produce a natural history of his voyage that Darwin later did. Instead, he eventually left his collections (herbaria, library, etc.) to Robert Brown. In 1896, Joseph Hooker edited and published the journal Banks kept of his voyage on H.M.S. *Endeavour*. Hooker remarked in his introduction that his principal motive in editing the journal kept by Banks was

---

[5] Patrick O'Brian. *Joseph Banks, A Life* (Boston: David R. Godine, 1993), p. 65.

[6] Lucile L. Brockway refers to Banks as the "unofficial founding father of Kew Gardens" in her comprehensive study of Kew, *Science and Colonial Expansion. The Role of the British Royal Botanic Gardens* (New York: Academic Press, 1979), p. 64. Brockway indicates that "the Royal Botanic Gardens at Kew, directed and staffed by eminent figures in the British scientific establishment, served as a control center which regulated the flow of botanical information." She indicates, "This botanic information was of great commercial importance, especially in regard to the tropical plantation crops, one of the sources of wealth of the Empire," p. 7. Toby and Will Musgrave in *An Empire of Plants* also refer to Banks as the "unofficial director of the Royal Gardens." They explain that because King George III and Banks shared a passion for plants, Banks was able to convince the King to have a place to research and record the entire flora of Britain's colonies, i.e., Kew Gardens, p. 149.

"to give prominence to" Banks's "indefatigable labors as an accomplished observer and ardent collector."[7]

When Captain Cook was informed that he would also be accompanied by "Mr. Banks and his Suite" a month prior to sailing, and that this entourage would be made up of nine persons, he understood that his naturalist was a man of influence. Banks was 25 at the time and obviously well connected, handsome, rich—he inherited his father's estate at 21—and intelligent—he attended Christ Church, Oxford University, although he never graduated. Banks viewed the voyage as an opportunity to further the study of natural history, especially botany, and convinced the Admiralty that he should accompany Cook at his own expense. He insisted on taking four servants, a secretary, two artists, a botanist, two dogs, and a great deal of baggage. Finding sufficient room on this relatively small ship was difficult and some of the junior officers, who were inconvenienced by this arrangement, were not pleased with this state of affairs.

Banks and his party, however, were largely responsible for the success of the voyage in view of the quantity of material in natural history and ethnology that was observed, collected, and studied, so the inconvenience of the junior officers never became much of an important issue afterwards. The extensive botanical collections by Banks and his botanist, Solander, would not have been assembled without both men's efforts. Banks's interest in taking artists resulted in an excellent visual record, establishing a precedent for taking artists on such voyages, and Alexander Buchan— who drew people and landscapes—and Sydney Parkinson (1745–1771)—who drew the plants and animals they collected—made rich contributions to the information gathered by Banks and his assistants. Buchan unfortunately died a few days after the *Endeavour* reached Tahiti so the responsibility of producing all the illustrations fell on Parkinson's shoulders, with some assistance from Banks's Swedish secretary, Herman Diedrich Spöring (1733–1771). In January, 1771, as the ship crossed the Indian Ocean on the way home, both Parkinson and Spöring died but not before Parkinson had accomplished a great deal of work. At the time he died of fever, Parkinson had completed 280 of the 900 drawings he had begun. When Banks and Solander returned to England, they worked with five other artists, and with the hundreds of dried and pressed specimens as a guide, to complete the project *Banks' Florilegium*. Of the completed watercolors, at least 743 were transferred to engraved copper plates by 18 skilled engravers. Banks sponsored this project by funding it himself.

But the project was never completed in Banks's lifetime. The work, in fact, was not published for two centuries. Solander passed away after Banks's return, and by the time the engraving was nearly complete (1784), Banks's income had declined, and publishing such a large book containing these elaborately colored plates would prove to be quite costly. Furthermore, Banks was now fully occupied as president of

---

[7] Joseph Dalton Hooker, *Journal of the Right Honourable Sir Joseph Banks made during Captain Cook's First Voyage on H.M.S. Endeavour in 1768–71 to Terra del Fuego, Otahite, New Zealand, Australia, the Dutch East Indies* (London and New York: Macmillan & Co. Ltd., 1896).

the Royal Society. The watercolors and sketches comprising the botanical record of Cook's voyage and the engraved copper plates and original specimens were eventually bequeathed by Banks to the British Museum (Natural History) where they have remained ever since (Fig. 2.1).[8]

Banks also directly influenced naturalists who did not have the opportunity to travel; for example, he assisted Erasmus Darwin (Charles Darwin's grandfather) by supporting Dr. Darwin's translation of Linnaeus's *Genera and Species Plantarum* and such sponsorship was Banks's primary role now because he did not serve on Cook's second voyage—on H.M.S. *Resolution*. Originally, he proposed to take a large party including naturalists, artists, servants, and even two horn players (tuba), and Cook agreed to give up his cabin and move to a renovated upper deck. When these changes made the ship unseaworthy, Banks threatened to withdraw his entire entourage. The Admiralty's patience with Banks finally was exhausted and Banks's withdrawal was accepted with some relief. Banks remained bitter about this episode, and a number of years later dictated his account of this episode to Brown.[9] Johann Reinhold Forster (1729–1798)—a man with whom Cook did not get on very well—was appointed naturalist for the voyage. However, accompanying the pugnacious naturalist was his precocious son Georg Forster (1754–1794). The presence of young Forster was a positive development because he was a gifted artist as well as a natural historian. He made many drawings of plants and animals observed or collected by his father. Banks purchased them in 1776 and today they are preserved in the British Museum (Natural History) in South Kensington, London (Fig. 2.2).

Forster's illustrations were an early source of inspiration for Alexander von Humboldt (1769–1859). Humboldt began his explorations in the Western Hemisphere at the beginning of the nineteenth century. His five-year expedition (1799–1804), traveling 6000 miles through remote rain forests and volcanic areas, helped establish the fields of plant geography, orography, and climatology. Humboldt's lectures on physical geography and related topics were subsequently

---

[8] The only twentieth-century printing, done in black ink, used 30 of the original plates and was published in a limited edition of 100 copies, in 1973. In the 1970s, the British firm Alecto decided to take on the task of printing from the surviving 738 original plates, the collection named *Banks' Florilegium*. The New York Botanical Garden has a set as well (set # 3). In addition to the vast botanical material—Banks and Solander had brought back over a thousand different species of plants unknown in Europe at the time, including 17,000 different individual plants—Banks gathered a wealth of information on animals. His lists of the arthropod species he collected while on the *Endeavour* were useful to early investigators. The botanical illustrations drawn by Parkinson supplemented the living material. Jonas Dryander's catalogue of drawings in Banks's library and Solander's notes on the voyage were additional help in sorting through the collections Banks brought back. Ann Botshon's brief summary of the fate of Banks's project is in her note, "Banks' Florilegium," *Garden*, November/December 1981, 14–17.

[9] Published in Edward Smith's *The Life of Sir Joseph Banks, President of the Royal Society with some notices of his friends and contemporaries* (London and New York: John Lane Company, 1911), pp. 25–26, f.n. 1. When Banks gave up the opportunity of serving as a naturalist on another voyage for Cook, he was a very arrogant and difficult young man Andrea Wulf observes in her study of the early pioneers of exploration in natural history, *The Brother Gardeners, Botany, Empire and the Birth of an Obsession* (New York: Alfred A. Knopf, 2009). She indicates that Banks over time mellowed and learned to curb his temper and be more diplomatic.

Fig. 2.1 Portrait of Sir Joseph Banks by Thomas Philips, about 1808–1809, Dixson Galleries, State Library of New South Wales, DG 25

published in a five-volume work titled *Cosmos* where he expressed his desire to "represent in one work the entire material universe," i.e., all phenomena and living things "that will stimulate the imagination."[10] The breadth of Humboldt's discoveries in all branches of the natural sciences was key in driving the development of botany in the Americas. Darwin and other Victorian naturalists often cited him as an important source of inspiration.[11]

Banks was not the original thinker that Darwin was. He was more of a descriptive scientist. This may explain why he did not go on to develop an evolutionary theory of his own or attempt to account for the great variety of life he observed. His letters reveal his limitations but also the powerful influence he had on his contemporaries.

---

[10] Alexander von Humboldt. *Cosmos. A Sketch of a Physical Description of the Universe.* Translated by Elise C. Otte, 2 vols. (Baltimore: Johns Hopkins University Press, 1997). Andrea Wulf indicates in *The Invention of Nature: Alexander von Humboldt's New World* (New York: Alfred A. Knopf, 2015) that Humboldt was a product of the German Enlightenment and investigated every natural component of the New World.

[11] Charles H. Smith indicates that Humboldt's written work not only inspired Alfred Russel Wallace to conduct voyages of exploration in natural history but also allowed him to embrace Humboldt's methods of carefully examining scientific phenomena in developing scientific laws. "Alfred Russel Wallace note 8: Wallace's earliest exposures to the writings of Alexander von Humboldt," *Archives of Natural History* 45(2) (2018): 366–369, 366.

Fig. 2.2 Caricature of Sir Joseph Banks, "The great South Sea caterpillar, transform'd into a Bath butterfly," James Gillray, desn. et fect., Pubd. by H. Humphrey, 1795 July 4th, Library of Congress Prints and Photographs Division Washington, DC, LC-USZC4-8780

In turn, his own development as a botanist was influenced by Linnaeus because Linnaeus was a dominant force in eighteenth-century science. This may account for the reluctance of Banks and others at the time to challenge scientific authorities by attempting to refute the idea that species were immutable or even Linnaeus's artificial system of classification that some found limited.

But Banks helped shape the development of botany, earth science, art, horticulture, and other aspects of English science and culture in the late eighteenth and early nineteenth centuries after his voyage on the *Endeavour*. He also advanced international cooperation in science by maintaining contact with France during the French Revolution and the Napoleonic Wars. All efforts to conduct travels on behalf of natural history in Britain went through him, and he had a role not only in the implementation of such projects but also in gathering the results of such journeys, the flora, fauna, fossils, and other geological samples, and seeds of many of the plants that were collected. His greatest legacy was the effect he had on the naturalists who followed him into a career of exploration in natural history, particularly Robert Brown.

Chapter 3
"… The Plants of Scotland Might Be Equally Useful"

Contents

Robert Brown's Early Life

How did Brown, a man of relatively humble beginnings, achieve his position as naturalist on a voyage of exploration? His adventure, promoted by the British Navy, enabled him to make a significant impact in natural history. Brown was born in Montrose, Scotland, December 21, 1773, located on Scotland's east coast (on the North Sea) into a family with a religious background, although it was somewhat unconventional.

He was the second and only surviving son of James Brown, an Episcopalian (Anglican) minister, and Helen Taylor, the daughter of a Presbyterian minister. At the time, Scotland imposed severe legal restrictions on Episcopalians (Anglicans). For example, James Brown could not address more than four people in a room at one time. He circumvented this restriction by preaching to his congregation in his home, structured in the shape of a cross. The congregation was spread out in the four rooms that defined the shape of the cross so that no more than four of the parishioners were in each room. James preached to this gathering from the stairs, and thus several more members of the congregation were accommodated in this "fifth room."

Young Brown was strongly influenced by his father's independent streak and ingenuity. He inherited his father's intellectual honesty, and grew to become quite self-reliant. In spite of the strong religious orientation of his parents, Robert showed little inclination to follow in their footsteps. When he began his formal education, his disinterest in clerical matters revealed a growing religious skepticism. He also developed a strong interest in natural history; the rugged and beautiful Scottish countryside may have been a factor in stimulating his fascination with the natural world.

© The Author(s), under exclusive license to Springer Nature Switzerland AG 2021
J. Schwartz, *Robert Brown and Mungo Park*, Memoirs of The New York
Botanical Garden 122, https://doi.org/10.1007/978-3-030-74859-3_3

Brown's experience at the local grammar school—today the Montrose Academy—was a positive one. There, he became friends with James Mill (1773–1836), a friendship that lasted until Mill's death. Mill was a utilitarian philosopher and economist but is best known today as the father of John Stuart Mill (1806–1873). Brown then went to Marischal College in Aberdeen in 1787. Marischal enjoyed a well-deserved reputation in developing botanists. The college's rich tradition of botanical studies helped produce Robert Morison who became Professor of Botany at Oxford University and David Skene who had corresponded with Linnaeus.[1] Brown benefited by his opportunity to study at Marischal College in Aberdeen. It reinforced his growing interest in natural history and allowed him to develop his inquisitive scientific mind, positive factors in his emergence as a naturalist.

Brown's name was on the students' registry at Marischal College until 1790 when he departed with his family for Edinburgh (a section known as Monteith's Close). His father's decision to move to Edinburgh was motivated by the chance to join a like-minded group of dissenters there who rejected the political decisions mandated by the Episcopalian leadership. The larger community in Edinburgh allowed the elder Brown to feel more secure.[2] He became pastor of like-minded renegades. James Brown's dissent from clerical matters bothered church leadership, but this did not prove to be a long-term problem for them because the elder Brown died in 1791.

Brown did not graduate from Marischal College because of his family's relocation. The move proved to be fortuitous nevertheless. Inspired by his new environment and the intellectual excitement of Edinburgh, he enrolled at Edinburgh University, taking medical courses beginning in 1790. Brown repeated this pattern throughout the period of his formal education, i.e., not completing his studies at one level before proceeding to the next one. Outwardly it suggests that he was not serious about his work but this is contrary to what he accomplished later on because he proved to be a very determined young man. Brown demonstrated unusual energy when something really interested him, pursuing the subject with tenacity. This determination applied particularly to botany as well as the entire field of natural history.

Brown attended Rev. Dr. John Walker's lectures in natural history at the University where Walker was Professor of Natural History and a horticulturalist. Brown collected plants for Walker, particularly specimens of bryophytes and lichens. In addition to collecting in the vicinity of Edinburgh, e.g., Brand Hill, Kings Park, and the ruins of Rosslyn Castle, Brown made a number of collecting trips to the Highlands during the summer months. Although Brown found Walker's lectures dull, he developed a good relationship with him. He received even more encouragement from another professor at the university, Daniel Rutherford (1749–1819), Professor of

---

[1] Marischal College, Aberdeen, enjoyed a rich scientific tradition. Later on, physicist James Clerk Maxwell taught there from 1856 to 1860.

[2] A freethinker, Brown's father was an independent man when it came to religious practices, forcing him ultimately to leave rural Scotland for Edinburgh.

Medicine and Botany and Regius Keeper of the Botanic Garden. Rutherford is best known for his discovery of nitrogen, an example of the multifaceted scientists Brown was exposed to at Edinburgh.

Rutherford had succeeded a friend and colleague of Walker's, John Hope (1725–1786). Hope studied in Paris with Bernard de Jussieu, an opponent of the Linnean system of classifying plants. This early exposure to different ways of looking at the taxonomy of plants had an impact on Brown's thinking. About the time Brown began his medical studies (1790), he may have encountered Mungo Park, a naturalist and a contemporary, also a dedicated botanist who, like Brown, had taken a greater interest in botany than medicine. But there is no evidence that the two men ever met or corresponded or had much to do with one another.[3]

Brown's contacts in Edinburgh were crucial to shaping his early career, particularly with such naturalists as George Don (1764–1814), James Dickson (1738–1822), and William Withering (1741–1799).[4] When he was only 17, Brown went with Don to Angus and discovered a previously undescribed sedge (*Scirpus hudsonianus*), a rush- or grasslike plant. Don was another example of the versatile naturalists who were helpful in furthering Brown's development as a naturalist in his own right. Don was a nurseryman (gardener) who specialized in the vegetation of the Scottish Highlands and later founded a botanic garden (at Doo Hillock in Forfar, Scotland). He was appointed Principal Gardener at the Royal Botanic Garden in Edinburgh in 1802.[5] Brown gathered a good deal of information from his fieldwork and this was the basis of his study, "The Botanical History of Angus," which was presented before the Edinburgh Natural History Society (on January 26, 1792). It was well received as were other works that followed. They were descriptive in nature, filled with information about the plants Brown discovered and collected during trips all over Scotland: the Edinburgh region and Ben Lawers, Glenlochar, and Ben Lomand in the western part; Dundee and Perth in the central regions of Scotland; Perthshire; Forfarshire; and the Highlands of Aberdeenshire.

While preparing his paper, Brown wrote to William Withering, a well-regarded botanist and physician, not an unusual combination for that period and characteristic of scientists who were the products of the Scottish Enlightenment. Withering

---

[3] David J. Mabberley speculated that Park might have introduced Brown to his brother-in-law, James Dickson, a Scot who had returned to Scotland after serving as nurseryman of Covent Garden because their interests were similar at the time, *Jupiter Botanicus, Robert Brown of the British Museum* (London: British Museum (Natural History), 1985), p. 19. Like Brown, Park also went on to reject Linnaeus's artificial system of classification based on the sexuality of plants but did not have the opportunity of further expanding his ideas on this subject. Brown and Park's careers did intersect later on, when Park indirectly, but significantly, impacted Brown's career by refusing to serve as naturalist on board H.M.S. *Investigator*, thereby opening the door for Brown.

[4] James Dickson, Mungo Park's brother-in-law, was a nurseryman who collected British plants and specialized in mosses. He opened a plant and seed shop in Covent Garden and became friendly with Sir Joseph Banks, introducing Park to Banks.

[5] Don was against church orthodoxy, but Brown's association with Don did not handicap his early career, and Brown, already influenced by his father's nonconformist religious views, became a religious skeptic.

also had received his education at Edinburgh University, and was perhaps best known for his use of an extract from the common or purple foxglove plant (*Digitalis purpurea*), initially utilized around 1785 in the treatment of heart failure. Withering also was a member of the Birmingham Lunar Society whose members included Erasmus Darwin (Charles Darwin's grandfather) and James Watt.[6] Brown reported the results of his botanical field studies, and asked Withering about some of the unusual plants he encountered. He offered to send samples of the plants he collected to Withering and asked him to send his reply to Rev. James Brown of Monteith's Close.[7] Withering replied promptly that he would be pleased to receive them and Brown sent them for use in Withering's herbarium. Brown knew about Withering because he was a friend of the Rev. William MacRitchie, a man whom he met on a field trip in the River Isla region. What is striking is the way these amateur botanists, many of whom were in the medical profession or clergy, accepted Brown, and encouraged him in his botanical work. Brown already was receiving recognition because of his keen eye and single-minded pursuit of unknown varieties and species. He sent them valuable samples of flora that were previously unknown in Britain.

With this encouragement, Brown continued his work on the flora of Scotland, methodically describing plants at the Royal Botanic Garden in Edinburgh. He also collected fungi in the Edinburgh area and other plants in nearby regions; for example, in Ben Bourde, he collected *Veronica alpina*, a member of the figwort family more common to the Pyrenees; the rare pea, *Oxytropis halleri*, which grows on inland or maritime cliffs and sand dunes; and a variety of unusual plants in the Breadalbane Mountains. He described these plants and sent samples on to Withering. Medicine had even less appeal now for Brown, and he discontinued his medical studies in 1793, because it had only served as a means to his obtaining a solid scientific education but now he had no further use of continuing his medical education. As a result he was freed to pursue his interest in botany and did so with great zeal. He continued to travel to the Scottish Lowlands as well as Highlands through 1794. Because his father died several years earlier in 1791, he no longer could continue to depend on his family for support.

Mungo Park

Mungo Park was born on September 10, 1771, in Foulshiels (Fowlshiels), four miles from Selkirk in the Royal Burgh of Selkirk, in the border region of Scotland. Mungo was the third son and seventh child out of 13 children, of the elder Mungo

---

[6] Withering was a progressive man of science, witnessed by his membership in the Lunar Society and his opposition to the phlogiston theory.

[7] There is a copy of the letter at the Birmingham, England Library; the original copy has not been found. See T. Whitmore Peck and K. Douglas Wilkinson's *William Withering of Birmingham* (Baltimore: The Williams and Wilkins Co., 1950) for a thorough account of Withering's life and contributions. What is interesting is that this volume does not discuss Brown at all.

Fig. 3.1 Portrait of Mungo
Park by Atkinson
Horsburgh, Peebles
Collection, Scottish
Borders Council,
PEEBM:6447

Park (1714–1793)—57 at the time of Mungo's birth—and Elspeth Hislop (1742–1817), who was 29. The elder Park was a prosperous tenant farmer on the estate of the Duke of Buccleuch. His father was able to provide a private tutor for his children, particularly the more extroverted Archibald, later the subject of Sir Walter Scott's novels, Alexander, to whom young Mungo was closest—he eventually became a lawyer—and Adam, who became a ship's surgeon. Although Mungo was more reserved than his siblings, he possessed a strong personality. He developed a love of Scottish ballads and stories and adventure. Later, he became a friend of Walter Scott, the great romantic novelist (Fig. 3.1).

Mungo attended grammar school in Selkirk, studying English, Latin, religion, mathematics, and geography. He was an excellent student and became interested in studying medicine despite his father's belief that he should prepare for the church. Like Brown's father, Park's father dissented from the views of his church, in his case the Church of Scotland. His dissent took the form of stricter adherence to religious orthodoxy than established beliefs, i.e., a devotion to pure Calvinist teaching—very much like the seventeenth-century Covenanters—and he joined a secessionist church in Selkirk.[8]

Young Mungo's friendship with Alexander Anderson at the school was a positive influence. Mungo's admiration for Alexander's father, Dr. Thomas Anderson, provided Mungo with additional motivation to study medicine. Mungo's father felt that

---

[8] The sect held beliefs close to those of the Puritans.

Mungo was headed for the life of a clergyman, believing that Mungo's mother, with her strong Calvinist beliefs, would influence his thinking and choice of a career. Instead, young Mungo went to live with Dr. Anderson at the age of 14 in Selkirk (in 1785), and served as an apprentice as he began the study of medicine. He often accompanied Anderson on his medical rounds and learned some pharmacology, i.e., the preparation (compounding) of drugs.

Park grew close to the entire Anderson family, particularly his contemporary, Alexander, who became his closest friend. In 1788, at the age of 17, he enrolled at Edinburgh University to begin formal medical studies. Alexander Anderson, more than a year younger than Park, joined him at the medical school in Edinburgh a year later. Park's brother, Alexander Park, joined Mungo and began medical studies as well. Park settled into Edinburgh life quite enthusiastically. Medical tuition was not excessive, and the University, with its excellent faculty including men the caliber of Joseph Black, a leading Professor of Chemistry, attracted many students to its medical school. A third of the entire student body of the University consisted of medical students. Park was particularly interested in botany, offered exclusively to students at the medical school. He was a dedicated student and exhibited some versatility by writing poetry and winning a poetry competition. The stimulating atmosphere of Edinburgh at the end of the eighteenth century, its more liberal religious beliefs together with the ideas of the Enlightenment, created an exciting environment for Park.

Park spent three years studying medicine at Edinburgh University (1788–1791), showing promise in his botanical studies as he toured the Highlands with his brother-in-law, James Dickson who had married his elder sister Margaret. Dickson was one of a number of Scottish botanist-gardeners who traveled south to England, particularly the London area, to work and study. Dickson worked as a gardener in several nurseries, and eventually opened an herb and seed shop in Covent Garden in 1772. He became friends with Joseph Banks, who became his patron as well. Banks allowed Dickson and eventually Park the use of his extensive library in London.

Dickson did not completely sever his ties with Scotland, and returned there on a number of botanizing trips where he discovered some plants not previously known to be native in Britain (e.g., snow gentian, *Gentiana nivalis*). His expertise focused on mosses as well as other flowerless plants (known collectively as cryptogams, for example, algae, fungi, and ferns, plants that reproduce by spores rather than by airborne pollen produced by flowers or cones; thus, the latter do not require moisture in reproduction the way cryptogams do). Dickson passed on his interest in cryptogams to his brother-in-law, Mungo (Fig. 3.2).

Park's relationship with Dickson eventually facilitated his introduction to Banks, already President of the Royal Society and a patron of exploration. Banks's experience as the naturalist on Captain Cook's first voyage around the world made him a firm advocate for journeys of exploration. His sponsorship of travel to distant and relatively unexplored lands and as a result returning with many observations along with examples of flora and fauna appealed to young promising naturalists as Park. Banks encouraged Dickson to continue his botanical research, allowing him the use of his library in Soho Square, London, the center of botanical research in England

Fig. 3.2 James Dickson,
F.L.S., by Henry Perronet
Briggs, 1820,
U.S. National Library
of Medicine

JAMES DICKSON F.L.S.

H.P.Briggs Esq.⟨ʳᵉ⟩ R.A pinx.ᵗ 1820.

at the time. When Dickson introduced Park to Banks, Banks was impressed by the younger man's knowledge of botany despite Park's initial shyness, and helped him secure a post on a ship that was to embark on a voyage of exploration. In addition, he helped Park gain membership, through Dickson's sponsorship, to the recently formed Linnean Society (1788) as an associate member. Dickson was a founding member of the Society.

Travels to Sumatra

In 1793, because of Banks's influence, Park was chosen to serve as assistant-surgeon or surgeon's mate on the *Worcester*, a vessel belonging to the East India Company's fleet of ships. While in Sumatra, a British possession run by the East India Company, Park would collect botanical specimens for Banks.[9] Park wrote to his good friend, Alexander Anderson, on January 23, 1793, informing him about his new position and also that he had purchased a copy of "[Dugald] Stewart's Philosophy [of the

---

[9] Banks received a letter dated 10 March 1792 from John Crisp (d. 1803), who was an official of the East India Company and a natural historian and fellow of the Royal Society. Crisp served as deputy governor at Fort Marlborough, Sumatra, at the time (1789–1793), and at the bottom of the letter there is a note in Banks's hand which he added to this letter: "Mʳ Mungo Park Surgeons mate Worcester," Sir Joseph Banks, *The Scientific Correspondence of Sir Joseph Banks* (London: Pickering & Chatto, 2007), volume 4, page 110, letter 1098.

Human Mind] to amuse me at sea."[10] Several weeks later (February 9, 1793), he wrote again to Anderson, revealing some trepidation because France declared war against Britain, realizing that the proposed journey to Sumatra contained some danger: "If I be deceived, may God alone put me right, for I would rather die in the delusion than wake to all the joys of earth."[11]

The *Worcester* had ten officers, including the Captain, John Hall, a crew of about 100, and a number of passengers and cargo. Park was registered as mate to the surgeon, Richard Lane.[12] Prior to his departure, Park was awarded £10 to pay for four months on board. He left from Gravesend on February 18, 1793.[13] Because Britain and France had then begun hostilities, the ship had to wait in Portsmouth until April 5th of that year so it could travel by convoy. Despite the outbreak of war, the journey proceeded without much hardship; there were few cases of serious illnesses or injury so Park was not overburdened by medical responsibilities. Park learned some elemental astronomy, enough to enable him to understand the methods of determining the ship's latitude and longitude. Captain Hall likely utilized John Harrison's method calculating longitude that had begun to be employed by the British Navy.

On August 22, 1793, the ship reached Bencoolen, Sumatra, which remained a Dutch possession although the British East India Company occupied the immediate area. The cargo was unloaded and new cargo was loaded for the return trip, consisting mainly of pepper, senna (an herb used as a purgative at the time), and an oriental spirit derived from coconut and/or palm sap called arrack. Park had time to dredge the waters, particularly along the shore, and when the ship moved from Bencoolen to Rat Island, he turned his attention to plant life while exploring the Sumatran mountains. His study of the fish he dredged led to a paper describing eight species of previously unnamed Sumatran fish, and it was first presented in a lecture to the Linnean Society when he returned in 1793.[14] When he returned home, Park gave Banks some of the plant specimens he had collected along with watercolors and anatomical descriptions of the species of fish observed at Bencoolen, as well as various rare Sumatran plants (Fig. 3.3).[15]

---

[10] A copy of the letter is in Joseph Thomson's *Mungo Park and the Niger* (London: G. Philip & Sons, 1890) [Argosy 1970 Reprint] pp. 42–43; the original copy has not been found.

[11] Thomson, *Mungo Park and the Niger*, p. 43.

[12] This information is provided in the *Worcester's* log, preserved in the National Maritime Museum, Greenwich.

[13] Kenneth Lupton's *Mungo Park, The African Traveler* (Oxford: Oxford University Press, 1979) established the correct year of the voyage by examining the log of the *Worcester*. It was a year later than reported by Park, who erred when he dated it a year earlier, and subsequent biographers continued the error until Lupton gave the correct date.

[14] The paper "Descriptions of Eight New Fishes from Sumatra" was published in *Transactions of the Linnean Society* (1797) 3: 33–38, and with the exception of its introduction was written in Latin. The paper was read before the Society on November 4, 1794.

[15] These watercolors are now housed in the British Museum of Natural History in South Kensington, London. Park's Linnean Society paper contains a beautifully colored chromolithograph of a species of *Perca*, *Perca canaliculatus*.

Fig. 3.3 *Perca lumulata.* A new species of Perch, collected and drawn by Mungo Park, Bencoolen, Sumatra, from "Descriptions of Eight New Fishes from Sumatra," *Transactions of the Linnean Society* 1797, Vol. 3, Issue 1:35–38

Park returned to Britain (May 2, 1794), imbued with a renewed interest in geographical exploration and natural history but further work along these lines would have to wait. He had to attend to family matters as well as deliver to Banks the plant specimens and his drawings and descriptions of Sumatran fishes. He also learned the distressing news that while he was on the *Worcester* (actually before the ship reached the equator in 1793), his father passed away at the age of 79. After he disembarked and collected his pay, he returned home to see his widowed mother and found that most of his siblings were away except for the youngest. With the money he earned from his duties on the *Worcester* and his inheritance, he felt financially secure. He now turned his attention to further exploration after making certain his mother was able to adapt to her new situation.

Chapter 4
Travels in the Interior Districts of Africa

Contents

Park's work in Sumatra had impressed Banks who was interested in promoting African exploration as a result of work with the African Association. This organization was founded originally as the Saturday's Club, consisting of 12 prominent men, aristocrats and members of parliament, a bishop, a retired general, and Banks, who served as treasurer. In 1794 Park offered his services to the African Association, then looking for a successor to Major Daniel Houghton, who had been sent in 1790 to discover the course of the Niger River and had died in the Sahara. Park wrote to his brother Alexander (May 20, 1794) informing him that he "got Sir Joseph's [Banks] word that if I wish to travel he will apply to the African Association and I am to hire a trader to go with me to Tombuctoo and back again."[1]

Park's First African Adventure

After resolving his affairs, Park went to London. The pay he received from his first voyage and his inheritance—his father left a decent enough sum for his children and wife—allowed him some financial security. Now he felt that he was ready for further adventure. The prospect of a "great river" that lay beyond the vast desert in northern Africa excited Europeans for many years. The death of Major Houghton,

---

[1] National Library of Scotland, MS. No. 107 82, f. 180. A copy is in Lupton, *Mungo Park, The African Traveler*, p. 17.

who attempted to navigate the Niger River on behalf of the African Association in 1791, was a stimulus for the group to find a new explorer. They had found their man with Park, whose scientific background was certainly more promising than that of Houghton who was merely a retired military officer. Banks regarded Park as uniquely qualified for such an expedition.

On May 22, 1795, Park sailed on the brig *Endeavour* from Portsmouth for the Gambia.[2] The African Association—Association for the Promotion of Discovery through the Interior of Africa—sponsored this expedition, after having been founded shortly before in the early 1790s. The Association was interested in exploring the African interior because of the failure of Houghton's expedition and in general the lack of exploration on the continent. Because some of the members of the organization were abolitionists and others were anti-abolitionists, they did not directly confront the issue of slavery in spite of the fact it was a burning question at the time. There were other concerns as well: economic, religious—the wish to conduct missionary work to convert Muslim and pagan tribes to Christianity—and scientific—interest in learning more about the natural history, geography, and ethnology of this territory.[3]

Park was instructed to get to the Niger by the most expeditious route and explore its course, from its point of origin to its terminus, visiting the major towns and cities located in the vicinity, especially Timbuctoo. Park described the circumstances that led to the expedition and what was expected of him:

> Soon after my return from the East Indies in 1793, having learnt that the Noblemen and Gentlemen, associated for the purpose of prosecuting Discoveries in the Interior of Africa, were desirous of engaging a person to explore that continent, by the way of the Gambia River, I took occasion, through means of the President of the Royal Society, to whom I had the honour to be known, of offering myself for that service. I had been informed, that a gentleman of the name of Houghton, a Captain in the army … had already sailed to the Gambia …. I had a passionate desire to examine into the productions of a country so little known …. The Committee of the Association, having made such inquiries as they thought necessary, declared themselves satisfied with the qualifications that I possessed …. Being favoured by the Secretary of the Association … with a recommendation of Dr. John Laidley (a gentleman who had resided many years at an English factory on the banks of the Gambia) …. My instructions were…to pass on to the river Niger … ascertain the course, and … the rise and termination of that river …. I should use my utmost exertions to visit the principal towns or cities in its neighbourhood, particularly Tombuctoo and Houssa (Fig. 4.1).[4]

---

[2] Because the *Endeavour* was a brig, and the ship Captain James Cook and Banks sailed on in 1769 was a barque it was unlikely the same vessel. It was common at the time to have a number of sailing ships bearing the same name. The *Endeavour* that Cook sailed on was sold by the Admiralty to the French Government in 1790 and was renamed the *La Liberté*. Its eventual fate is unknown although there are reports that it lies at the bottom of Newport Harbor in Rhode Island while a contrary report indicated that years ago it was at anchor in the Thames, between Woolwich and Greenwich, having spent its last days as a home for female convicts. *Cook's Log* (Published by Captain Cook Society), vol. 19, no. 2 (1996), p. 1273.

[3] Kenneth Lupton in *Mungo Park, The African Traveler* indicated that Banks was interested in the continent's rich natural resources, including gold, but "the botanist … was also interested in the natural products growing, or capable of being introduced and grown …" p. 25.

[4] Park, *Travels in the Interior District of Africa* (London: W. Bulmer and Co., 1799, Third Edition), pp. 1–3.

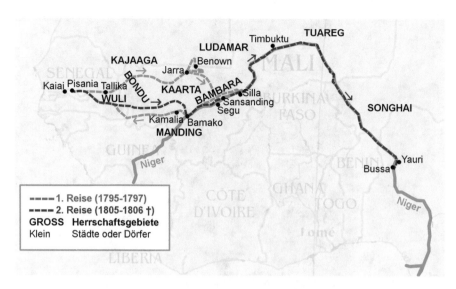

Fig. 4.1 Map depicting Mungo Park's Trips to Africa, created by Jens Nachtigall, nachtigall@web.de

Park and his party arrived at Jillifica in the Kingdom of Barra on the northern bank of the Gambia on June 21, 1795. Park remained on the *Endeavour* as it continued going up the Gambia to Jonkakonda where he disembarked. He went overland and then traveled on the Gambia River for 200 miles to Pisania, a village founded by British and other Europeans for the purpose of trading slaves, ivory, and gold. With a letter of introduction from the African Association, Park met Dr. John Laidley, the owner of a small British factory, who warmly welcomed him. Laidley treated Park quite hospitably and he used the time he spent there as Laidley's guest learning the Mandingo language—with Laidley's assistance—the language in general use in this region of Africa.

While in Pisania, he contracted a fever and suffered delirium. This set him back for months so he was unable to start out for nearly a half a year. So his planned expedition into the interior was delayed. However, he was able to use the time to his advantage while recovering. He observed the different tribes of people who lived on the banks of the Gambia with whom he came in contact: the Feloops, the Jaloffs, the Foulahs, and the Mandingoes. He tried to learn the rudiments of their language and discovered that iron was the European product that interested them the most. He also found that in agriculture, the plough was unknown as a tool and donkeys served as "beasts of burden" (most commonly referred to as "asses").

The African slave merchants were called *Slatees*. They supplied the inhabitants of "maritime districts" with iron, sweet-smelling gums, and frankincense, plus *Sheatoudou*, which essentially was tree butter, a commodity extracted by means of boiling in water the kernel of a nut of the shea or vegetable butter tree (*Vitellaria*

Fig. 4.2 Shea or butter tree from *Mungo Park's Reise in Afrika für die Jugend* (1805)

paradoxa).[5] During his stay in Pisania, Park also studied some of the plants found in the area. He sketched and made watercolors of them including the leguminous *Erythrina senegalensis*, commonly known as the coral tree and referred to as *nté* in West Africa, where it is still used medicinally (Fig. 4.2).[6]

Park Search in the Interior

Park finally set out from Pisania on December 2, 1795, leaving the comforts of Laidley's mansion behind. To act as a guide and interpreter, he was given an African servant, Johnson, who had lived in Britain for some time and who had a good grasp

---

[5] Park, *Travels in the Interior District of Africa*, p. 26.

[6] Park's sketchbook, containing the drawings and watercolors he made while in Pisania, is preserved today in the Library of the Royal Botanic Garden in Edinburgh.

of not only the Mandingo language but English as well. He also took a horse for himself, two donkeys for servants, provisions for several days, a small assortment of beads and trinkets, amber, tobacco, some changes of clothing, an umbrella, a pocket sextant, a magnetic compass, a thermometer, two fowling pieces (including a light shotgun used for killing game), and two pairs of pistols. Laidley was pessimistic about Park's chances of success, fearing he would not see him again but Park was undaunted as he headed toward Medina, the capital of Wooli. He learned that the Mandingoes were divided into two sects, the Mahomedans, called *Bushreens*, and the Pagans who were called *Kafirs* (meaning unbelievers) or *Sonakies* (i.e., men who drink strong liquors), with the Pagans the most numerous by far.[7]

Arriving in Medina, the capital of Wooli on December 6th, the king there welcomed Park and his party. The king, described in several accounts as benevolent, warned Park not to travel into the interior because Major Houghton had been killed when he traveled on the same route. Furthermore, white men were not known in these territories.[8] Nevertheless, Park set out toward Bondou, passing through Tambacunda (December 9th), Koujar (the frontier town of Wooli on the 11th), where the inhabitants greeted his party "with a mixture of curiosity and reverence."[9] He also noted a large tree, called by the natives *Neema Taba* that was decorated with many rags or scraps of cloth, possibly a religious practice. They entered the frontier town of Tallika in Bondou, a state that is the present-day Senegal, and at the time was inhabited chiefly by Foulahs of the "Mahomedan" religion. The native people there lived in "considerable affluence."

Bondou lay between the Gambia and the Senegal rivers, and it was better drained than the territory he had left behind. Park found the land quite fertile, indicated by the abundance of mimosa trees. Park and his party left Tallika on December 14th. They crossed a large branch of the Gambia called the Neriko where the banks were steep and covered with mimosas. On December 18th, they left the village of Doogi and then went on to Soobrudooka. Park noted that the banks of the Falemé River were covered with large and beautiful fields of corn. The natives called it *Manio* (named *Bolcus cernus* by botanical writers), learning that it grew during the dry season. Park observed that the further east he traveled, the climate became drier, and the landscape was more interesting in that it seemed more varied.

About noon on December 21st, Park and his party entered Fatteconda, the capital of Bondou. He received an invitation to the house of a respectable *Slatee*. The chief, who previously had ill treated Major Houghton, made Park trade gold for his blue coat with its gilt buttons. Park agreed to the chief's demands to avoid any difficulties. The party was able to depart without further complications or problems because of Park's acquiescence.

They then entered the Kingdom of Kajaaga, traveling at night to avoid thieves as well as the dense woods that were filled with wolves and hyenas. The inhabitants

---

[7] Park, *Travels in the Interior District of Africa*, p. 34.

[8] *Ibid.*, p. 37.

[9] *Ibid.*, p. 41.

there were referred to as "Serawoollies," called "Seracolets" by earlier French travelers. There was a serious setback when in the frontier town of Kajaaga, Joag, the king's son, took half of Park's possessions on the pretext he had entered the kingdom without permission. A nephew of the king of the adjoining territory of Kasson happened to be visiting and offered to escort Park to his uncle's capital. Because Park literally had to beg for food at this point and because his newly found escort's return to his home was on the same route he intended to take, Park accepted the offer.

Park reached the town of Samee on the banks of the Senegal River at the end of 1795. Upon his arrival, Park's self-appointed guide left him, taking more of what few possessions remained. On December 28th he departed from Samee and arrived in the afternoon at Kayee. He was detained in Teesee because while there, in the territory of Kasson, "a handsome present" was requested. This deadlock lasted for almost 2 weeks until it was determined that he would be of no further use and was allowed to depart Teesee on the morning of January 10, 1796, for the Kingdom of Kasson. On the morning of the 15th he was granted an audience with the king of Kasson (Demba Sego Jalla). He explained to the king the reasons for his journey and reported in his journal that the king appeared satisfied with his explanation.[10]

It was at that point Park decided to change the easterly route (actually ESE) he initially believed would take him to the Niger. He decided to proceed on a more northerly route to Ludamar. On February 1st he learned that war had not yet begun between the Bambara and Kaarta and felt that he could pass through Kaarta before the Bambara invaded. His servant Johnson advised him that the route Park had chosen was quite dangerous, so much so that Johnson, although his trusted guide and servant, decided to return to Pisania where they had begun their hazardous journey. Thus Park and Johnson parted company in Jarra, in the Kingdom of Ludamar—a predominantly Islamic state—but not before Park entrusted Johnson with a copy of the journal he had kept thus far. The following day after Johnson's departure, Park came upon some natives gathering the fruit of *Rhamnus lotus*, which he discovered was converted into a type of bread that was an important food for the natives of Kaarta and Ludamar.

Park's Capture

At this point, Park was nearly out of provisions so he set out for the camp of Ali at Benown, ruler of Ludamar—described as the Moorish king of Ludamar—accompanied only by Dr. Laidley's slave boy and a messenger from Ali who promised to assist him in his passage to Benown. However, despite traveling at night to avoid the "Moorish banditti," Park was robbed and beaten by fanatical Moors, who looked upon him with particular hatred because he was regarded as a Christian infidel. He was to receive much worse treatment yet at the hands of Ali, an unusually cruel tyrant.

---

[10] Park, *Travels in the Interior District of Africa*, p. 85.

Before they parted, Johnson had pleaded with Park not to proceed further toward Ludamar because he feared the Moors and the punishment they dished out. Park's difficulties only confirmed Johnson's worst fears. Because Park allowed Johnson to return to the Gambia carrying his journal—keeping a duplicate for himself—he had decided not to put anyone else at risk, perhaps indicating that he was aware of the dangers that awaited him. When Park was captured by Ali (March 7, 1797), he was further humiliated while imprisoned in Ali's camp in Benown. Ali ordered his beard shaved, and took the rest of his possessions except his pocket compass, so that he would not run away. Ali was largely indifferent to Park, keeping him imprisoned with little food and no protection from fanatical mobs that intended to inflict great harm. In addition to the dangers associated with starvation, there was also the mental strain of not knowing if and when his life would be taken.

Perhaps he was spared because the early part of his captivity corresponded with the month of Ramadan. At first, this did not seem a positive development because he was denied food and water during the Muslim period of fasting but despite the constant humiliation he suffered, his captors seemed to be unable or not ready to make a decision as to Park's eventual fate. Time seemed to work in his favor because he also benefited from the attention of Queen Fatima, Ali's chief wife. Fatima took pity on this strange intruder perhaps out of kindness. She sent him food and drink. One of the indignities he suffered was that he was not allowed to drink water from the vessels that the Moors drank from but often had to drink from a trough that had been constructed for that purpose. Despite his feverish and perilous state, Park tried to remain as calm as possible and keep his sanity and bearings. He was aware of the civil war being waged around him. He looked for a possibility of escape.

Before Park was able to escape, Johnson was captured in Jarra and brought to Benown. Johnson had the rest of Park's clothing and other personal possessions such as his watch and other instruments but fortunately Johnson had secured Park's journal papers with the wife of one of his protectors so this important record continued to remain in safe hands. Although now united, both Park and Johnson remained captive (Fig. 4.3).

At the end of April, there was news of an attack by Mansong, a former ally of Ali who was angry with Ali's failure to come to Mansong's aid in his fight against Kaarta. So the entire camp moved to Bubaker where Ali spent time with his wife Fatima. Conditions were terrible there with a serious shortage of water and Park's servant Demba begged for provisions and water. Ali declared his intention to make Demba his slave and Park was given the choice of either Demba or himself.

At the beginning of June, Park saw Daman Jumma in Jarra while he was still captive, quite a different circumstance from when they first met. In spite of his reduced state, Park was able to extract the promise from Daman that Demba would be eventually freed. He and Demba had to part but Park was told he could take Johnson. His situation further improved when Ali went back north to continue his battles with the Kaartans, leaving Park and Johnson behind in Jarra. Daman treated Park and Johnson fairly because of a promise he previously had made to Laidley. By the end of June, Desse, another warlord, was expected to arrive with his army so many fled in panic. After four months, Park saw an opportunity of escape. While Ali

Fig. 4.3 A view of Ali's Tent at the Camp of Benown, where Park was held captive by Ali in 1797, from *Mungo Park's Reise in Afrika für die Jugend* (1805)

was distracted by an attack from a former ally, Park would be able to flee but without the rest of his possessions, except his compass and horse. Also, he would no longer have his servant Demba who had become Ali's personal slave.

Park's Escape

Park learned from Johnson that Ali's chief slave and other Moors had come to take Park back to Ali. This information made Park more resolved than ever to risk escaping. Park and Johnson made their bid for freedom during the night of July 1st while their captors were asleep. Johnson expressed his desire to return to the Gambia taking Park's papers with him while Park was determined to push on toward the Niger and fulfill his mission. There were some more anxious moments when he was set upon by a group of Moors who also declared their intention to take him back to Ali. However, they were more interested in robbing Park, leaving him alone after taking a cloak that protected him from the elements. In July of 1796, he was at last free.[11]

With the aid of his compass, Park was able to set a course, east-southeast, south of the route he had taken going across Ludamar, prior to his capture. Barely able to

---

[11] Lupton speculated that Ali might have lost interest in keeping Park any longer and made a rather half-hearted attempt to continue holding him in captivity. It is difficult to determine the accuracy of this account, and Park's memoirs (*Travels*) do not provide a definitive answer, *Mungo Park, the African Traveler*, p. 71.

slake his thirst or give his weary horse enough water to keep going in the arid conditions of Ludamar, Park finally reached the greener territory of Segu. He received
better treatment in villages along his route, receiving enough food and drink so he
was able to go at a fairly brisk pace, averaging close to 30 miles a day. The rainy
season had begun, and he frequently had the company of refugee Kaartans, hoping
that could further secure help when reaching Segu. His tattered and pitiable appearance was the cause of some amusement but he was no longer in any danger. At a
small village, Park learned from the locals that the Niger was nearby (called the
Joliba, meaning "the great water"). On or about July 21st, 1796, Park arrived at the
Niger, the great object of his journey:

> As we approached the town, I was fortunate enough to overtake the fugitive Kaartans to
> whose kindness I had been much indebted in my journey through Bambara. They readily
> agreed to introduce me to the king; and we rode together through some marshy ground ...
> one of them called out, *geo affili* (see the water), and looking forward, I saw the great object
> of my mission—the long sought for majestic Niger, glistening to the morning sun, as broad
> as the Thames at Westminster, and flowing *slowly toward the eastward* (Fig. 4.4).[12]

Segu, the capital of Bambara, was a relatively prosperous kingdom at the time,
situated in the south central part of present-day Mali. It consisted of four distinct
villages, two located north of the Niger, and the other two to the south. Park reported
that the "numerous canoes on the river, the crowded population, and the cultivated
state of the surrounding country, formed altogether a prospect of civilization and
magnificence which I little expected to find in the bosom of Africa."[13] However, this
did Park little good as he was a pitiable sight, and quite hungry. Mansong Diarra,
ruler of the Bambara Empire, refused to see him lest to incur the wrath of the
Moorish population. Instead, he was told to go to one of the villages in the area
where he was turned away from every door when he asked for food and rest: "The
king could not possibly see me, until he knew what had brought me into his country;
and I must not presume to cross the river without the king's permission."[14] One
person, a woman who had returned from her "labours in the fields"—probably picking cotton—had pity on him and provided both shelter and food; he enjoyed "a very
fine fish," as a result.[15] All Park could give in return were two of the four brass buttons from his waistcoat that he had reacquired, possibly after his reunion with
Johnson earlier in the year.

The king inquired through messages whether Park had anything for him and Park
informed him that he had been robbed leaving him quite destitute. Another messenger informed Park that Mansong ordered him to leave but "wishing to relieve a

---

[12] Mungo Park, *Travels in the Interior Districts of Africa* (Ware, Hertfordshire: Wordsworth
Editions Limited, 2002), pp. 178–79. Reprint of *Travels in the Interior Districts of Africa* (London:
W. Bulmer and Co., 1799) with additional material, p. 194. Park indicated that it was on the 20th
when he first saw the Niger. Modern accounts such as Lupton's in *Mungo Park, the African
Traveler* have calculated it more accurately, i.e., it was the 21st of July.

[13] *Ibid.*, p. 180.

[14] *Ibid.*

[15] *Ibid.*, p. 182.

Fig. 4.4 Mungo Park seeing the Niger at Sego for the first time in 1796 (detail), from Sir Harry Johnston's *Pioneers of West Africa* (Blackie, 1912), Look and Learn Ltd

white man in distress" provided him with 5000 cowries (little shells used as currency), which would allow him to purchase provisions on his journey along the Niger and hire a guide. Park speculated that Mansong would have given him an audience but was fearful that he would not be able to protect him "against the blind and inveterate malice of the Moorish inhabitants." Park concluded that the king's behavior was "prudent and liberal" but thought that the king believed Park had concealed the true purpose of his journey.[16]

Park proceeded down the Niger along the northern bank, likely by horseback and foot. He narrowly escaped attack by "a large red lion." He first went to Sansanding, then to Moodiboo—where he enjoyed a wonderful view of the Niger for miles in either direction—Moorzan, and finally arrived in Silla without his horse, which was quite broken down before he lost possession of it. In Silla, he confirmed what the natives had told him—quite contrary to what he believed before he left Britain—that the Niger flowed eastward, not westward, and learned that his ultimate destination, Jenné, was two days of travel eastward, and beyond that he learned that the

---

[16] *Ibid.*, p. 183.

Niger passed through a large lake, Dibbie (Debo). Also, the further east he traveled, the Bambara language was less familiar to the people he encountered. Park, relying on secondhand accounts of the region's geography, did not receive entirely accurate information. After learning that another tribe of people (the Maniana) were cannibals, and because he already was in a feverish state, he decided to return to the region of Gambia and then home (July 30, 1796):

> Worn down by sickness, exhausted with hunger and fatigue; half-naked, and without any article of value by which I might procure provisions, clothes, or lodgings ... I was convinced ... that the obstacles to my further progress were insurmountable. The tropical rains were already set in with all their violence ... in a few days more, travelling of every kind, unless by water, would be completely obstructed. The cowries which remained of the king of Bambara's present [Mansong] were not sufficient to enable me to hire a canoe for any great distance and I had but little hopes of subsisting by charity in a country where the Moors have such influence. But above all, I perceived that I was advancing more and more within the power of those powerful fanatics; and from my reception, both at Sego and Sansanding I was apprehensive in attempting to reach even Jenné .... In returning to the Gambia, a journey on foot of many hundred miles presented itself to my contemplation, through regions and countries unknown. Nevertheless, this seemed to be the only alternative, for I saw inevitable destruction.

Park, in an almost apologetic tone added, "With this conviction on my mind, I hope my readers will acknowledge that I did right in going no further."[17]

When Park decided to reverse course, he knew he would encounter great difficulty on the return trip. He had a few articles of clothing and the rest of his possessions were gone, and the tropical rains had set in. Although travel was impossible except by river, he did not have a sufficient amount of cowries to obtain a canoe. His supply of the shells given to him earlier by Mansong had run low. He set out nevertheless, often wading through three feet of water, and encountered hostile Moors. Furthermore, the natives grew suspicious of him because of Mansong's refusal to see him. He decided to avoid villages as much as possible, and follow the recommendation of the Dooty (the "chief man") of a less hostile tribe and travel westward on the northern bank of the river. The Dooty allowed him to speak with his fishermen and they took him to Moorzan. Here, he hired a canoe for 60 cowries and the Dooty's brother, who was traveling to Modiboo, allowed Park to accompany him, and even was accommodating enough to carry Park's saddle that Park had been able to keep.

When he arrived in Modiboo he learned that his horse had been recovered although the poor animal was not in the best of shape. Thus he would be able to travel on horseback again. When he reached Nyamee, the heavy rains made it impossible to travel immediately on the course he set. He departed finally for Sibity, where the Dooty informed him that his people were otherwise engaged, and he asked for a guide to take him to Sansanding. In Sansanding, he was not greeted warmly as he had been before because he learned that a very "unfavourable report" had been received from Segu about him. Park was also informed that Mansong had

---

[17] *Ibid.*, p. 195.

Fig. 4.5 A view of the village of Ramalia where Park regained his health in 1797–1798, from *Mungo Park's Reise in Afrika für die Jugend* (1805)

dispatched a canoe to take him back so it seemed prudent to leave the Segu area as quickly as possible.

Toward the end of August, he encountered a group of armed people, the Fulahs, an African tribe of unknown origin who had lighter skin than most of the native inhabitants. They took most of his clothes, his horse, and the rest of his belongings, leaving him with a rather worn shirt, pants, and a top hat with its crown where he kept his notes. These notes, containing geographical information, possibly were saved because he had secured them in his hat. His assailants were wary of taking further chances, fearing that Park possessed too much magic to remove his hat. Just when he seemed to have little hope of survival, fortune seemed to smile on him as he dragged himself to the frontier town of Manding, Sibidooloo. The Dooty seemed friendly and promised that his property would be retrieved. On September 8, 1796, his horse and clothes were restored to him but his compass was broken and was of no further use. His horse fell into a well and Park, realizing that the animal was in a skeletal state due to insufficient food, presented him to the Dooty who had treated Park with kindness. The only request Park made was to treat the horse well. He sent his saddle and bridle to the king of Sibidooloo, who was instrumental in recovering his stolen property.

While in Sibidooloo, he came down with a severe fever, probably malaria, and he struggled on to Kamalia. There, his fortunes took a turn for the better. Karfa Taura, a slave merchant, took him to his own home, and helped care for him until he recovered. He continued to suffer from fever throughout the rest of the rainy season. Park stayed as Karfa Taura's guest for seven months of until he regained his health, and he did not ask Park for any compensation for the food and shelter he received (Fig. 4.5).

Park's Travels Home

After the rains subsided, Park felt a bit better but quite weak. On April 19, 1797, he left with Karfa Taura—after waiting until his benefactor was ready—to undertake the trip to Pisania, a distance of 500 miles through thickly wooded hills separated by many rivers and streams, a trip that was not as difficult as the trials he faced earlier but mentally was very trying. Karfa Taura was taking with him a caravan of slaves, who were prisoners of war and wrapped in leg irons, and driven by the lash. They were carrying heavy loads on their heads and were forced to march 30 miles a day.

Park reached Pisania in June 1797 and was greeted warmly by the Westerners there because they had given him up for dead. He was able to change into the clothes he had left at Laidley's house and shave off his beard, looking once again like a Westerner. He arranged with Laidley to pay for Johnson's wages and also left some money to see if Demba's freedom could be restored although he had not learned of his fate for some time.

Park booked passage on the first ship that was available, an American vessel, the *Charleston*, bound for South Carolina, with a human cargo of slaves. This was not the most direct route back to Britain but there were not too many opportunities to arrange any sort of passage. Before crossing the Atlantic, the ship, captained by Charles Harris, sailed up the river to pick up provisions at Gorée. There was a considerable delay in picking up supplies to cross the Atlantic. Because Gorée was a French fort and Britain was at war with France, Park might have been at risk, but because the *Charleston* was an American ship, the *Charleston* finally was allowed to proceed across the Atlantic during the first week of October 1797.

Conditions on board were crowded, and the ship's surgeon had died from fever in the Gambia so Park agreed to act as the ship's surgeon. This was much appreciated because as well as ministering to the ill, Park knew enough of language from his years in the interior of Africa so he could communicate with the slaves. Although he did not witness excessive cruelty, he observed that the conditions on American slave ships were worse than on British ones. Over 20 of the original 130 slaves perished by the end of the journey. The ship never reached South Carolina as the condition of the ship was so poor it landed in Antigua in the West Indies. This trip took five weeks but in 10 days after arriving in Antigua, Park was able to secure passage to England on the mail packet, the *Chesterfield*. Upon arriving in Antigua in November, he wrote to Banks (a letter dated November 13, 1797), informing him of the circumstances surrounding his unexpected detour to Antigua, and explained that because of this unanticipated stop in Antigua, he needed an additional amount from Banks totaling £40.[18]

---

[18] Park wrote, "The vessel in which I had taken a passage for America was obliged to come in distress, and I have this day taken a passage in the Chesterfields packet for England, in consequence of which I have been under the necessity of Drawing on you in favour of Sherrington & Dixon, for the sum of forty Pounds ...." Letter from Park to Banks, Antigua, November 13, 1797, in the collection of copies of the correspondence of Sir Joseph Banks made for Dawson Turner, Fellow of the Royal Society in 1833–1834. *Dawson Turner Collection* (*D.T.C.*) 10 (1): 209.

The voyage returning home went relatively smoothly certainly compared to his previous experiences, welcome relief after the trials he has been subjected to. He reached Falmouth in Devon on December 22, 1797, after being away from England for over two and a half years (actually two years and seven months). People had given him up for dead or at least lost with little hope of survival. When Park arrived in England, he wept with joy and gratitude for his deliverance, and arrived fittingly in London on Christmas day. *The London Chronicle* for Friday, January 5, 1798, reported, "Mungo Park, the African traveler, returned to London a few days ago, having penetrated into the interior of Africa, much farther than any of his predecessors; and not less than twenty-two degrees Eastward from the Cape de Verd Islands."[19] Other periodicals were equally jubilant and celebrated Park upon his return.

Park did not document many observations in natural history because he was unable to collect many specimens of the unusual plants he observed. Also, he frequently was handicapped because often he had no means of transport. In spite of the many obstacles he faced, Park brought home approximately 80 plant specimens including two that are named after him, the shea tree, *Butyrospermum paradoxum*, with the subspecies *parkii* and the locust bean tree, *Parkia biglobosa*. Even so, with the exception of descriptions of plant life in Park's *Travels*, particularly as it related to the native foods and the animal life such as lions and elephants, this work primarily was concerned with the different native peoples he encountered, their language, their customs, and their religion.

As a result, this journey could be viewed as an early attempt to study ethnography, much as later travelers such as Thomas Henry Huxley made along the shores of Australia and New Guinea. More importantly, his adventures and discovery of the course of the Niger, but not its entire path, fired the imagination of members of the African Association, including Banks.[20] After, when Banks offered him the opportunity to travel to Australia, it seemed quite proper recognition of his talents and reward for his exploits. Those plans did not materialize.

Park's Celebrated Return

Park was greeted as a hero when he returned to Britain. The African Association finally had a live explorer to honor, and he became much celebrated in London society. Park met with Banks and related the details of his expedition and shared with him the botanical specimens he was able to bring back with him from Africa, meagre as they were. Reports in the magazines and other periodicals provide testimony

---

[19] *The London Chronicle*, from Thursday January 4 to Saturday, January 6, 1798, No. 6061, p. 17.

[20] The Niger River, as Park observed, flows eastward for a considerable distance before reversing direction, and then it travels in a southwesterly direction to the Atlantic. See Glyn Williams, *Naturalists at Sea, From Dampier to Darwin* (New Haven and London: Yale University Press, 2013). p. 6.

to the way the British public viewed his accomplishments. *The London Chronicle* was one such example of the periodicals heralding Park. The entire notice (in the Friday, January 5, 1798 issue) reported:

> Mungo Park, the African traveler, returned to London a few days ago, having penetrated into the interior of Africa, much farther than any of his predecessors; and not less than twenty-two degrees Eastward from the Cape de Verd Islands. Among other interesting notices, his discoveries concerning Lotophagi [legendary people who ate lotus and encouraged travelers to do the same], confirm the account of that singular people given by the Greek writers.[21]

The same periodical contained a lengthier notice several weeks later. Some of the details, including the spelling of Park's name and that Park had been to Houssa, were incorrect but the account gives a good idea of how celebrated a figure Park had become. The story from Saturday, January 27th, related:

> Mungo Parke, the gentleman employed by the Members of the African Association, to explore the interior of Africa, arrived in London some few days ago, with, we are credibly informed, the whole of his papers and observations, which are of the most important nature to this country. To satisfy public curiosity, an abstract of his journal will shortly be published, preparatory to the more voluminous part of his travels.
>
> Mr. Parke, we understand, fully confirm the account given by Major Houghton of the population and extend of the city of Houssa, which lies on the great river near Tombuctoo. The unfortunate and premature death of the Major is likewise confirmed beyond all doubt. Indeed, Mr. Parke himself narrowly escaped a similar fate. From the very extraordinary discoveries of Mr. Parke, the British merchant may anticipate vast sources of wealth. This traveler, a native of Scotland, has penetrated farther into Africa than any European living; and for his successful enterprise, his country will not doubt amply reward him. The inhabitants of the city of Houssa, twice as large and twice as populous as London, are exceedingly partial to those specimens of the English manufacture which Mr. Parke presented to them. Our woolens, linens, Manchester fluffs, and hardware, will find there a safe and ready market, uninterrupted by any rival Power; and we will receive in return gold and other precious metals, with which the country round Houssa naturally abounds, besides many other productions which may be rendered highly interesting to us.[22]

The St. James's Chronicle (*British Evening Post*) also reported Park's discoveries in celebrating his return from Africa. Under a postscript in the paper dated, "London: Thursday [January 4, 1798] One' O'clock," it repeated the account in *The London Chronicle* that Mungo Park, the African traveler, returned to London a few days ago, penetrating into the interior of Africa, further than previous explorers, 22 degree east from *Cape de Verd* Islands. It added, "He brought with him accurate delineations of his travels; and is preparing for the press a full account of his interesting journey."[23]

---

[21] *The London Chronicle* from Thursday, January 4, to Saturday, January 6, 1798, Volume LXXXIII, No. 6061, Thursday, January 5, 1798, p. 17c.

[22] *The London Chronicle* from Thursday, January 25, to Saturday, January 27, 1798, Volume LXXXIII, No. 6070, Saturday, January 27, 1798, p. 96c.

[23] *St. James's Chronicle* (*British Evening Post*) No. 6247, Tuesday January 2 to Thursday January 4, 1798, January 4 (p. 4, last page).

The main daily newspaper at the time *The True Briton*, on Monday January 29, 1798, contained a detailed and accurate description of Park's adventures:

> Among the discoveries of Mr. Mungo Park, the African traveler, one of the most important is, that the River Niger runs Eastward, as is said by Herodotus. It is in most places larger than the Thames, and is navigated by double canoes. Some of the Kingdoms near its banks extend two hundred miles in length, and near half as much breadth. Mr. Park travelled near two thousand miles inland, from the Western coast of the ocean. He found the inhabitants Negroes, with a slight mixture of Moors. They cultivate the ground, using hoes. They do not make use of ploughs or cattle. Their manufactures are cloth of cotton, which every family weaves for his own use. They have iron ore, which they fuse with charcoal, and make of it knives, &c. Their towns are meaner than those of South Bombay; the houses of one story, flat roofs, and unadorned by any public monuments. Through almost the whole extent of Mr. Park's travels, he found the lotus, which affords a farinaceous substance made into bread, and which, with the Indian corn, is the chief support of the inhabitants. There is another tree which he calls the butter tree, because the kernels of its nuts afford a substance exactly resembling butter in its taste, as well as colour and consistency. The natives are ignorant Pagans; physic [art of medicine] and conjuring are the most useful trades in travelling through their country. Their medium of exchange consists in gold dust, and the shells called Cowries which pass as coin also in India. They are not cruel or unfriendly to strangers. A King of one of the largest districts, about fifteen hundred miles from the Western Coast, though he suspected Mr. P. to be sent as a Spy, yet dismissed him with a present of 5000 cowries. Lions and other wild beasts are not formidable obstacles to a traveller. Mr. Park was more afraid of meeting with one Moor than with twenty Lions.[24]

Aftermath of Park's Journey

After his return, Park resumed his life in Selkirk. Initially on his return, he devoted himself to writing *Travels in the Interior Districts of Africa* (1799), where he reported his observations and explained that he believed that the Niger flowed eastward, speculating that it eventually veered south and flowed into the Congo River.

Meanwhile, Park turned down the government's offer (and Banks's as well) to serve as surgeon and naturalist for an expedition to explore Australia. While this was happening, Park married Allison Anderson, the daughter of the man he had been apprenticed to. Instead of accepting the government's offer by traveling to Australia, Park practiced medicine in Peebles (the town was already renowned because it was where the illustrious self-made publishers and writers, William and Robert Chambers, were born and spent their early years).

Many have ascribed Park's refusal to agree to go on the Australian expedition to his marriage to Allison Anderson but that does not seem plausible enough in light of the much more demanding and dangerous trip he undertook less than a decade later (in 1805–1806), when he left not only his wife but also young children, as well as a medical practice in Scotland to return to Africa. A better explanation seems to be due to a combination of factors, particularly the less than magnanimous attitude (or

---

[24] *The True Briton*, No. 1591, Monday. January 29, 1798 (p. 4, last page of issue).

penurious treatment) on the part of the Admiralty, underlined by the meagre pay they promised him as well as his desire to return to Africa to complete the job he felt was necessary. Park was insulted by the government's offer. On September 14, 1798, he wrote to Banks that his initial reaction to Banks's "intended letter" to the Duke of Portland recommending him for the expedition to New South Wales made him feel like he had achieved his highest ambition particularly because he "heard with pleasure that the Association had recommended my services to [the] Government." But subsequently he realized "the wage indeed appeared too small." In addition when he learned from Governor King that the "ship drops down the river in the course of ten days … it will be impossible for me to make the necessary preparations."[25]

Park followed up his letter of September 14th with another one to Banks (on September 20th), further explaining the reasons for his refusal. He thanked Banks for what he did, writing that he "had considered at leisure the subject of our conversation at Spring Grove [Banks's country estate] and tho I have every reason to be satisfied with the part you have taken in the business, I feel that those trifling misunderstandings have considerably damped that enthusiasm … necessary to ensure the success of such an enterprise; and, as I have no wish to deceive, nor disappoint my patrons by engaging in an enterprise which from the circumstances I mentioned to you cannot be perfectly agreeable–I must … wait for some other opportunity, when perhaps I may be better reconciled to the situation."[26]

Banks replied a day later (September 21st) that only Park was "acquainted with the real Motives of your now declining an Engagement … which you always appear'd to me to solicit with Eagerness," and that Park alone was "able to determine … the propriety of your Conduct." Banks indicated that he would not offer any arguments but from his previous transactions with Park he was surprised to find his enthusiasm damped by the small difference in the pay expected and offered.[27]

Apparently Banks was not ready to let the matter rest despite his comments. He wrote to Robert Moss (d. 1801) Under-Secretary of State at the time, on September 21, 1798, and explained that "he [Park] is determin'd not to go to New Holland," because he expected better compensation although Banks speculated that that might not be the entire story. He speculated, "I have since more reason to suspect is really the case. At all Events as he is a Free Man we cannot compel him …. I had form'd to myself a great Idea of the discoveries he wou'd have made, which will not now

[25] Letter from Mungo Park to Banks, September 14, 1798 [in a collection of autograph letters compiled by Kenneth A. Webster] Webster Collection <13>; *D.T.C.* 11: 71–72. Also copy in *Sir Joseph Banks, The Indian & Pacific Correspondence of Sir Joseph Banks*, ed. Neil Chambers, vol. 4, letter #348 (London: Pickering & Chatto, 2011), pp. 548–549.

[26] Letter from Park to Banks, September 20, 1798, *D.T.C.* 11: 79. Also, copy in *Sir Joseph Banks, The Indian & Pacific Correspondence of Sir Joseph Banks*, ed. Neil Chambers, vol. 4, letter #350 (London: Pickering & Chatto, 2011), pp. 549–550.

[27] Letter from Banks to Park, September 21, 1798, *D.T.C.* 11: 80. Also, copy in *Sir Joseph Banks, The Indian & Pacific Correspondence of Sir Joseph Banks*, ed. Neil Chambers, vol. 4, letter #352 (London: Pickering & Chatto, 2011), p. 551.

be explor'd in my Life time."[28] Moss responded to Park (on September 24th) by increasing his offer and expressed the hope that "Park's final determination will be in the affirmative; as it will be much to be lamented if the undertaking should be given up."[29]

Banks informed Park (on September 25th) that he "received a letter form the Secretary of State's office authorizing him to offer 12s. 6d. a day, 2s. 6d. more than originally fixed; respecting outfit, he will undertake that he receives a sum sufficient for all necessary instruments and for presents to the natives; as he conceived all misunderstanding had been explained in the interview he had with him, he is surprised to find him pleading it a second time as a reason for declining the engagement; misunderstanding must now cease and he must state openly his motive for declining."[30] Park, however, displayed the stubbornness that formed much of his character. He replied (on September 26th) that "12s. per day and 100 pounds for outfit was originally proposed but he afterwards found that only 10s. was offered and no outfit; he regrets he cannot he cannot accept the terms offered even as they now stand."[31]

However, this affront or what Park believed to be a rather insulting offer was not the only reason why he turned down the offer to travel to Australia as a naturalist. Park actually supplied a motive for his action in his lengthy account of his first trip to Africa, *Travels*. In this work, he apologized to readers for not quite completing his mission, and the regrets he expressed there could be regarded as evidence that he wished to return to Africa and complete the task he began on his first expedition there. Park, in other words, despite the acclaim he received, did not regard his first journey as an unalloyed success.

The prospect of traveling to Australia, on perhaps a safer or at least less dangerous journey, presented greater opportunities for a naturalist. However, Park viewed it as a diversion from what he felt was his primary task: i.e., to finish the job he began on his first expedition to Africa and complete tracing the course of the Niger. Adventure and redemption seemed to weigh more heavily than the opportunity to do significant work in natural history. Thus, Brown became the eventual beneficiary—not Park—because Brown was the one who eventually went on to make significant contributions in natural history.

---

[28] Letter from Banks to Robert Moss, September 21, 1798, copy in *Sir Joseph Banks, The Indian & Pacific Correspondence of Sir Joseph Banks*, ed. Neil Chambers, vol. 4, letter #351 (London: Pickering & Chatto, 2011). p. 550.

[29] Letter from Moss to Banks, September 24, 1798, *The Indian & Pacific Correspondence*, letter # 354, p. 552.

[30] Letter from Banks to Park, September 25, 1798, *D.T.C.* 11: 86–87. Also, copy in *The Indian & Pacific Correspondence* letter # 355, pp. 552–553.

[31] Letter from Park to Banks, September 26, 1798, *D.T.C.* 11: 88–89. Also copy in *The Indian & Pacific Correspondence*, letter # 356, p. 553. Privately, Banks was a good deal harsher in criticizing Park's rejection. He wrote to Moss on September 28th (1798), explaining that if Park "had his Enthusiasm continued, he wou'd have made an excellent Instrument in the hands of a good Director, as that is gone, he is no longer worth a farthing," *The Indian & Pacific Correspondence*, letter # 357, p. 554.

Although Park's pride was wounded, resulting in his refusal to go even though the government eventually raised the ante, it did not permanently destroy his relationship with Banks. Initially, relations became strained between the two men but eventually they were resolved, because Banks was a forgiving sort who was loyal to the protégées he selected; once he believed in someone, he was confident in his initial judgment and he knew how difficult it was to negotiate with the government over this issue. The fact that Park was engaged to marry at the time he was offered the position of naturalist to travel to Australia may have presented Park with an additional reason for refusing the opportunity but the most important consideration was that Park wanted to return to Africa to complete what he initially had undertaken. The mystery of what happened to the Niger after Timbuktu was an unsettled matter, and Park wanted to play a part in solving this puzzle.[32]

---

[32] Park's marriage is the reason often cited as why Park refused Banks's offer. But Park's marriage never impeded his daring adventures later on so it seems that it was not the chief reason for his refusal. Mark Duffill, in his fascinating biography, *Mungo Park, West African Explorer*, makes it clear that it was Park's hurt pride that led him to turn down the opportunity Banks presented him. But even this explanation is insufficient. Park turned down Banks because he wanted to return to Africa to continue the exploration of the Niger, something he was unable to do on his first trip there (Edinburgh: Royal Museum (National Museum of Scotland), 1999).

Chapter 5
A Good Practical Botanist

Contents

The Fencibles in Ireland

In spite of not finishing his medical education, Brown joined the army because he could draw on a dependable source of income and be financially independent to pursue his interests in natural history. He was accepted into the newly established regiment of the Fifeshire Fencibles. They were formed to quell the restive conditions in Ireland.[1] Commissioned on October 20, 1794, Brown was sent to Ireland with the rank of Ensign and Assistant Surgeon, having the necessary background despite not finishing medical school. Britain had been at war with France since 1793 so his service in Ireland served as a respite. He continued his botanical work, and with the added bonus of conducting his studies in a new locale. When Brown joined his regiment, he found that his new situation matched his expectations; Ireland proved to be a new territory to explore and collect its native plants. Luckily for Brown, the Fifeshire Fencibles saw little action, but his medical training gave him additional status. On June 1, 1795, he was appointed Surgeon's mate.[2] He carried out his duties while pursuing his interest in botany, and the military found him a diligent officer, so much so that they sent him to London to recruit. This proved to be fortuitous because it was on such a trip he met Sir Joseph Banks, in 1798.

---

[1] The Fencibles were the only full-time regulars who were limited to home service (i.e., "defencible"), unless all members voted to go overseas. Thirty-four Fencible Cavalry regiments were raised in 1794–1795.

[2] Brown wrote to Withering later—on January 29, 1797—that little or nothing has occurred. David J. Mabberley, *Jupiter Botanicus, Robert Brown of the British Museum* (London: British Museum (Natural History), 1985), p. 31.

© The Author(s), under exclusive license to Springer Nature Switzerland AG 2021
J. Schwartz, *Robert Brown and Mungo Park*, Memoirs of The New York
Botanical Garden 122, https://doi.org/10.1007/978-3-030-74859-3_5

It was during this period (1795–1798) that Brown undertook a remarkable program of self-education, learning German and expanding his personal library of botanical works and medical texts.[3] He continued his contact with Withering by post regarding his progress in collecting. His trips to England also allowed him to deliver samples of his Irish collections but he was unable to see Withering personally. Many of the specimens he collected he had sent to Withering for the latter's herbarium. Unfortunately during this period, Withering had become ill, so his son, William Jr., who was studying at Edinburgh University, began to assist his father. As a result, Brown enjoyed greater contact with William Jr. and had not yet met the distinguished botanist, the senior Withering.[4]

Brown's study regimen increased in intensity. He studied the plants he gathered even more acutely. The only instrument he used in his examination of the fine structure of the plants was a simple microscope, developed by the celebrated British naturalist, John Ellis. Ellis designed it to study small aquatic organisms; thus it was given the name the "John Ellis aquatic microscope." Brown became quite adept in his use of the microscope, and maintained this skill throughout the rest of his career. He later imparted this skill to others including young Charles Darwin before the latter embarked on his voyage of discovery on the *Beagle*.

Brown's Extended Stay in Britain

Brown took advantage of a sick leave and traveled to Edinburgh in late 1796. He did further work on Scottish plants, thereby receiving additional recognition. When conditions in Ireland grew more difficult he returned there. However, the increase in hostilities eventually led to his being sent to England for an extended stay from 1798 to 1799 in order to recruit for the military, which was desperate for new manpower. Britain was already at war with France and now was faced with a growing rebellion in Ireland. Brown was grateful for this respite. He used his extended stay to visit the oldest botanical gardens at Oxford—where there was no one to greet him and he was rather lost among the stodgy atmosphere of that University town—and the Chelsea Physic Garden, a more productive visit where he met with people he had been in communication with for some time. His meeting with the ailing Withering took place at the latter's home at Edgbaston. It proved to be quite fruitful, and Brown was delighted in finally seeing the man he communicated with so frequently. He showed Withering the Irish plants he collected, including a pipewort

---

[3] Later on, Brown was able to utilize these language skills he had acquired earlier. K.A. Austin in *The Voyage of the Investigator 1801–1803, Commander Matthew Flinders, R.N.* (Adelaide: Rigby Limited, 1964) indicated that Brown's linguistic abilities helped Commander Flinders; "Flinders could not speak French and for this reason he took Brown, an accomplished linguist with him" when the party from the *Investigator* met with Captain Baudin of the French fleet, p. 108.

[4] After William Sr.'s death in 1799, Brown helped complete the 1801 edition of *Botanical Arrangement*.

(*Eriocaulon aquaticum*), a beautiful white-flowered plant that grows in shallow waters that Withering had described earlier.

During this period, Brown was elected as an Associate to the Linnean Society of London, quite an honor for someone relatively young (25). In addition to visiting the Society in its rooms in Panton Square three different times during the year, he worked in the herbarium of the British Museum, and the herbarium of Edward Forster (1765–1849) at Walthamstow to collect more plants and add to the lists of plants he had assembled.[5]

Banks' librarian, Jonas Dryander (1748–1810), secretary to the Linnean Society, was one of Brown's sponsors for membership, and James Dickson served as the other. Equally important was his visit to the herbarium and library of Banks at Banks's home in Soho Square in London, then the center of British botany. Dryander, impressed with Brown's body of work completed so far, facilitated Brown's gaining ready access to Banks's collection. At the time, Banks' residence in Soho Square was the finest botanical collection in all of Britain. It had developed from Banks's own voyage of exploration (on Captain Cook's first voyage), on the bark, H.M.S. *Endeavour* more than 20 years before (1768–1771), and the collection expanded with the work of many other naturalists who sent the foreign specimens they collected on their expeditions to Banks. Brown was too reticent and uncomfortable because of his innate shyness to introduce himself to Banks. The meeting had to be deferred until later on, when an emissary helped Brown with a proper introduction.

Brown's reticence and caution, despite the considerable skill he already demonstrated and the accomplishments he had already achieved, are perplexing, epitomized by his reluctance to meet with Banks directly. Initially, he wrote to Withering for an introduction so he could work at Banks's herbarium and library even though a letter of introduction was probably not required in these circumstances.[6] Certainly, his reputation should have facilitated a proper meeting as well as guarantee access. Withering eventually wrote to the Portuguese émigré and political refugee, José Francisco Correia da Serra, on Brown's behalf. In the meantime, Brown visited many herbaria and gardens, particularly in the vicinity of London, assiduously describing and collecting flora native to the area, meeting and widening the circle of naturalists he knew, particularly Dryander. He spent a good deal of time in the early autumn at Soho Square researching plants from the genus *Eriocaulon*, which fascinated him, and at the British Museum, examining the notes and collections of Daniel Solander, who had accompanied Banks on Captain Cook's first voyage on the *Endeavour*.

In October 1798 during his recruiting mission for the Fencibles in London, Brown met Abbé José Francisco Correia da Serra. Correia da Serra was a Portuguese

---

[5] Phyllis I. Edwards reported that he visited the Linnean Society on February 6th, 1798, and July 5th and 17th of the same year, "Robert Brown (1773–1858) and the natural history of Matthew Flinders' voyage in H.M.S. Investigator, 1801–1805," *Journal of the Society for the Bibliography of Natural History* 7 (1976): 385–407, 387.

[6] Collection at the *Royal Gardens at Kew*, Letter from Brown to Withering, June 15th, 1798.

politician and scientist as well as an especially skilled botanist who upset the authorities in the church as well as his government in Portugal. Correia da Serra found it necessary to flee to England where he fell under the protection of Banks. He joined the Royal Society and found comfortable refuge there as secretary in the Portuguese embassy in London. Later on, he had a quarrel with the Ambassador and went to Paris, and eventually settled in America.

When Brown met Correia da Serra, the latter was impressed by the young man's determination and dedication, and Correia da Serra realized that Brown should meet Banks. Brown's meeting with Correia da Serra occurred at an opportune time, not only for Brown but also for Banks because soon he would need a naturalist for a voyage of exploration that was several years in the making and had been cancelled and then modified. Correia da Serra wrote to Banks on October 17, 1798, at Banks's home in Soho Square while the latter was staying at his country estate at Revesby Abbey in Lincolnshire, explaining that Brown was "a very good naturalist who frequents your library where I have made acquaintance with him, hearing that Mungo Park does not intend to go any more to New Holland offers to go in his place. Science is a gainer in this change of man."[7]

Brown spent some time with Correia da Serra, working at the Royal Society and the Linnean Society. They got on very well and Correia da Serra again noted young Brown's skills and dedication. The exiled naturalist felt that Sir Joseph must meet this impressive and dedicated young scientist. It took some time for Banks to get to know Brown because Brown's natural diffidence made him reluctant to put himself forward and approach Banks directly. So Correia da Serra performed a valuable service in taking the trouble to make Banks aware of Brown. At the time, Banks was continuing his search for a naturalist who would accompany Matthew Flinders on a voyage to New Holland, very much a matter of concern not only to naturalists like Banks but also to the British Government interested in this territory as a potential replacement to the colony Britain had lost in America.

When Brown finally met Banks at his home in Soho Square, London (sometime after Correia da Serra's letter of introduction to Banks in October 1798), Banks was already faced with a dilemma. His first choice was to have Park go as naturalist on the journey to New Holland (Australia) to explore the interior of the continent, and collect as much flora and fauna as possible. But Mungo Park refused to go and Banks still needed a naturalist to go on the planned journey. Banks still harbored the faint hope that Park would change his mind. However, due to the clumsiness on the part of the government and Park's easily bruised feelings and the latter's desire to eventually go back to Africa and complete the exploration of the Niger River, there was no chance of this occurring.

---

[7] Correia da Serra letter of Oct. 17, 1798, to Banks, *Kew Gardens, Banks Correspondence.* 2. 206; also a copy in Edward Smith's *The Life of Sir Joseph Banks, President of the Royal Society, with some notices of his friends and contemporaries* (London and New York: John Lane, 1911), p. 231, and Sir Joseph Banks, *The Indian and Pacific Correspondence of Sir Joseph Banks* ed. Neil Chambers (London: Pickering & Chatto, 2012, vol. 5, Letter [4]), p. 4.

The Plan to Explore New Holland

Originally, the expedition was planned differently and was to be conducted on a different vessel, the *Porpoise*. Banks's idea was to send Park to explore the interior of the continent, and a somewhat inexperienced botanist from Strangeways, near Manchester, George Caley (1770–1829), to collect specimens. Caley had written to Banks for several years, imploring him for a position. He had sent Banks some specimens of mosses, encouraged to do so by Manchester botanists he worked with as well as Withering.[8] Banks was not entirely pleased with Caley, and urged him to work at the Chelsea Garden or at Kew to get further experience with non-domestic plants, warning him that no trade was "so unremunerative as Botany."[9]

Caley was not satisfied with Banks's brush-off, and indicated more than once that he remained interested in making a career in botany. He asked for a role in the expedition to New South Wales but Banks continued to insist that he should work at Kew to increase his knowledge. On January 7, 1798, Banks wrote to Caley, "If you are dissatisfied here & chuse [sic] to go back to Manchester I can have no objection."[10] On July 12th, Caley wrote to Banks brusquely that by not continuing at Kew, he knew that he had made Banks angry but if he could not go to Botany Bay, he "could not live on the wages paid at Kew." He explained that he was too "fond of botany to take up more profitable work."[11] Obviously ambitious and eager for adventure, Caley was reluctant to follow Banks's suggested course of action even though James Dickson advised him that to regain Banks's favor he should return to work at Kew.[12] But an "unexpected opportunity" arose; Banks received permission from Governor Philip Gidley King (1758–1808), who was taking over responsibilities of governing the territory of New South Wales, for Caley to go with King to New South Wales.[13] King had a checkered and somewhat controversial career. As a former governor of the notorious penal colony at Norfolk Island, he had less than a favorable reputation but he had sponsored natural history studies in the areas under his jurisdiction.

---

[8] Caley was the son of a Manchester horse dealer. David Mabberley speculated that Withering's involvement with Caley may have prevented him from writing to Banks earlier on Brown's behalf, *Jupiter Botanicus*, p. 40.

[9] This remark is from a letter from Banks to Caley, March 7, 1795, *Dawson Turner Collection* (*D.T.C.*) 9: 199–200. Brown later praised Caley's collecting work. He referred to him as "*Botanicus peritus et accuratus*," high praise indeed from someone with Brown's meticulous standards. Also see Alice M. Coats, *The Plant Hunters: Being A History of the Horticultural Pioneers, Their Quest and their Discoveries, From the Renaissance to the Twentieth Century* (New York: McGraw-Hill Book Company, 1969).

[10] Letter from Banks to Caley, January 7, 1798, *D.T.C.* 10 (2): 157–158.

[11] Letter from Caley to Banks, July 12, 1798, *D.T.C.* 11: 6–8.

[12] Caley's letter of August 23rd 1798 to Banks indicated that he was dissatisfied with his "treatment" and gave Banks an ultimatum that if his letter "is not answered within 10 days," he will assume that Banks is not acting in a proper manner. *D.T.C.* 11: 37–43.

[13] Banks' letter to Caley, November, 16, 1798, copy in British Museum (Natural History), *D.T.C.* 11. 116–117.

Most of these ambitious plans fell apart when Park decided not to go on the major part of the expedition. When Brown eventually was chosen as a replacement, Caley refused to join the expedition but instead, despite Banks's trepidations about Caley and his abilities, Caley basically traveled on his own to New South Wales. He left on the *Porpoise* but it proved to be unseaworthy so Governor King and Caley were then transported on the *Speedy* and they reached Port Jackson (now Sydney) in April 1800. Later on, Caley assisted Brown on a number of collecting trips in Australia, particularly to the Blue Mountains, and later on, Mount Hunter.

Banks was more upset about the fact that the other part of the planned expedition fell through. In May 1798, Park refused the proposal that he undertake exploration in the Australian interior. Initially, Banks continued to believe that Park was eager to go because Park's main objection at the time was that the proposed remuneration on the part of the government was inadequate. When Park wrote to Banks to inform him of his dissatisfaction, Banks promised Park that he would try to raise the offer and had him see the Under-Secretary of State in September of 1798. Banks thought that would resolve the matter. However, the government thought that the pay was sufficient and they refused to budge. Banks became quite upset with both parties, and again tried to get the government to raise their bid but it was too late in any event. The Admiralty's displeasure with Banks may have been due, in part, to the earlier episode concerning Banks's preparations to go as the naturalist on Captain Cook's second voyage on H.M.S. *Resolution* and it reminded them of Bank's excessive demands.[14] The problem also was that during this period of time, the government was weak and racked with indecision.

Brown's Opportunity

Park's refusal to take part in the expedition opened the door for Brown. This proved very fortunate in the long run because Brown, although perhaps not as flamboyant a figure as Mungo Park, was probably a far more diligent observer of nature. In spite of health problems that occasionally cropped up, Brown was indefatigable in his labors. When Correia da Serra wrote to Banks on October 17, 1798, recommending Brown to fill the position of naturalist, he enthusiastically commented, "Science is the gainer in this change of man," adding, "Mr. Brown being a professional naturalist" to emphasize the point.[15] When Brown finally met Banks, he made a very

---

[14] Although Cook very much wanted Banks to be part of the much-anticipated journey because of the success of the prior voyage on *Endeavour*, he was ready to relinquish his cabin to accommodate the large party Banks planned to take along, but the British Government's patience had run out. The Admiralty was annoyed with Banks's plans to take on so many supernumeraries—including tuba players—believing it as excessive. They firmly refused to accept Banks's conditions despite his connections in society.

[15] Correia Da Serra letter of October 17, 1798, to Banks, *Kew Gardens, Banks Correspondence*. 2. 206; copy in *Correspondence of the Rt. Hon. Sir Joseph Banks Bart*, Botany Library of the British Museum (Natural History) 11. 111; in Edward Smith's *The Life of Sir Joseph Banks,* p. 231, as well as in *The Indian & Pacific Correspondence*, vol. 5, letter # 4, pp. 4–5.

favorable impression upon the latter, and his formal appointment as naturalist for the proposed voyage took place at the end of 1800. What was significant was that Banks recognized Brown's abilities at their initial meeting. He observed that Brown would tenaciously work on a problem until he was satisfied with its resolution.

Banks was also fortunate with his eventual choice of Matthew Flinders (1774–1814) to be in charge of the ship, *Investigator*, in conducting the expedition in the waters surrounding New Holland. Flinders, commander of H.M.S. *Investigator*, who is still known as the "man who first circumnavigated Australia" (a.k.a. "the Great Denominator"), is credited also with establishing the name of the continent, Australia. Like Brown and unlike Banks, he had a modest upbringing and came from a family with little nautical experience. He was a good student and was always fascinated by the sea. He received training and some experience but his boldness in writing to Banks when he was just a junior lieutenant in September 1800 seems on the surface to be rather brazen. However, Flinders had the talent and success in previously charting the territory near Van Diemen's Land (Tasmania) and the nearby Bass Strait during a voyage around the world on H.M.S. *Providence* under the command of Captain William Bligh. Flinders's experience was an additional factor in allowing Banks to look upon him favorably. Flinders was also responsible for inventing a device for counteracting the magnetic interference on a compass produced by the ship, the "Flinders bar." This device was a soft vertical iron bar that was placed on the foreside of a compass binnacle. It was used to counteract the vertical magnetism produced by the ship, thereby correcting this deviation by a process known commonly as "swinging the compass." (Fig. 5.1)[16]

When Flinders offered his services to Banks to command the expedition, Banks felt that he had found the right man to take on this task. However, Flinders was faced with a difficult problem and the prospects for the journey; H.M.S. *Investigator* was not in the best of shape, and unfortunately this eventually would pose serious problems. The ship, a "sloop-of-war," that was converted from a collier, was not properly caulked due to rotting wood under its copper-sheathed hull. Thus it leaked. Luckily, Flinders enjoyed the services of a good although rather small crew (Fig. 5.2).

It took some time for all these things to fall into place. In late 1798, prior to being officially appointed as naturalist, Brown traveled to Bath to do recruiting, primarily the reason why he was in England. Bath was an interesting place to do recruiting. This fashionable resort was in the grip of fear of French invasion.[17] Furthermore, Brown felt that too often unqualified men had been taken on to serve so he and his superior officer's had hoped to straighten that out by getting better men to serve. While committed to this task, Brown found time to write to his mentor, Withering, informing him of his work on *Eriocaulon* (pipewort) and other botanical

---

[16] See Matthew Flinders, "Concerning the Differences in the Magnetic Needle, on Board the Investigator, Arising from an Alteration in the Direction of the Ship's Head," *Philosophical Transactions of the Royal Society* 95: 186–197 (1805).

[17] Fear was so intense that the year before the poets William Wordsworth and Samuel Taylor Coleridge as well as their wives were under suspicion of being French agents probably because of Wordsworth's sympathies with the French Revolution.

Fig. 5.1 Portrait of
Matthew Flinders around
the time he was Captain of
the *Investigator* by Antoine
Toussaint de Chazal
Chamerel, about
1806–1807, Art Gallery of
South Australia, 20005P22

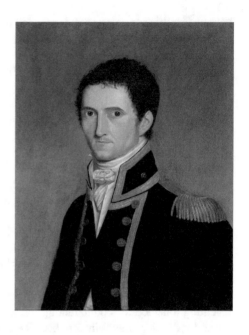

HMS Investigator, commanded by Matthew Flinders - July 1802

Fig. 5.2 (center) *H.M.S.* Investigator, *commanded by Matthew Flinders—July 1802* by John
Sheard Grafton (detail), 2006, from Elynor Frances Olijnyk, AHA AAMH, *Taking Possession: A
Saga of the Great South Land* (Marino, South Australia: Photographic Art Gallery, 2007),
©Photographic Art Gallery, 2007

observations, and he asked for specimens of cryptogams (lower plants such as fungi, algae, liverworts, mosses, and lichens).

Recruiting records for December 1798 mention that Brown had some success with three new recruits but they were less than ideal, and he made further efforts to enlist whatever qualified men could be found in the Birmingham area.[18] Brown complained of an infection in his cheek that had bothered him for some time, and when he recovered, he requested to be sent to London. In May of 1799, Brown was happy to be in London again, buying books for Withering, and working at the British Museum. This line of work was quite productive in that he was able to identify new genera, some of which he named and others included the moonwort, *Botrychium*. He identified a group of "tree ferns" belonging to the genus *Nelsonia* (named after David Nelson, the gardener at Kew who sailed with Cook). He also visited Banks's herbarium and library.

About this time, Brown was corresponding with Sir James Edward Smith (1759–1828), also a Scottish botanist and founder of the Linnean Society. Very much like Brown although a decade older, he exhibited a keen interest in the natural world at an early age, and similar to Brown, Smith enrolled in medical studies at Edinburgh University where he studied chemistry with Professor Joseph Black and natural history with Professor John Walker. He moved on to London in 1783 to continue his work and became friendly with Banks. Smith had acquired a good deal of the library and specimens of Linnaeus, which he offered to Banks, but inexplicably Banks refused the generous offer. Smith made good use of this collection when he founded the Linnean Society in 1788 and became its first President, remaining so until his death in 1828.[19] Brown's purpose in writing to Smith was to get advice from him regarding the papers he was preparing, a natural thing for him to do since the subject was often examples of species that their professors at Edinburgh, e.g., Walker and Hope, had studied (Fig. 5.3).

Smith was the author of one of several works concerned with the subject of Australian flora published before the voyage of the *Investigator*. One work, *A Specimen of the Botany of New Holland*, was published in 1783–1795. It contained information regarding its flora and fauna including watercolors, sent by John White (1756–1832), who collected specimens in his spare time while he served as Surgeon General in Botany Bay, 1788–1795. The other was Jacques J. H. de Labillardière's *Novae Hollandiae Plantarum Specimen*, which contained the results of his own

---

[18] "Recruiting records" in James R. Sutherland's, *Oxford book of literary anecdotes* (Oxford University Press, 1975), pp. 159–163.

[19] When Smith died, the Society was able to purchase Smith's personal collection along with what the Society did not own, forming the crux of its collection. A notice in *Gentleman's Magazine* (1828) 143: 296–300 indicated that when Smith arrived in London, "he became acquainted with Sir Joseph Banks, that eminent patron of natural science ... upon whose recommendation purchased in 1784 the celebrated Linnean collection .... Having purchased the Linnean collection, and settled in London as a man of acknowledged science ... [he] in conjunction with Dr. Goodenough, Lord Bishop of Carlisle ... and Thomas Marsham, esq ... set about establishing the Linnean Society of which Dr. Smith was the original President, and to which distinguished office was annually and unanimously chosen, from that period to the present time," 297.

collecting and those of David Nelson. Labillardière (1755–1834) studied medicine and botany, and had also corresponded with Banks and Smith, adding further observations in natural history.

Brown eventually returned to Ireland in 1799 where he resumed his military duties involved with treating the sick and wounded. It is interesting to note that despite a lack of a medical degree, his medical training was deemed sufficient enough to allow someone with his background to treat the wounded and those who were ill. Part of his military responsibilities was his attendance at court martial trials, mainly minor offenses like theft. He seemed to be treading water at this point although he busied himself keeping up with the work of other botanists, and working on his botanical observations and reading. In August, he was posted to Maynooth in the Dublin area. Later, he was posted to Kilcock and then to New Abbey and Kilcullen Bridge, which 2 years before was the setting for some of the worst of the Irish Rebellions. Happily for Brown, the Aberdeen Regiment replaced the Fifeshire so he was spared the rather intense and fierce fighting there. He enjoyed the comradeship of his fellow officers like his friend, George Kerr, who sometimes went on botany field trips with him. His letters were filled with his botanical observations, and at the beginning of 1800, he resumed teaching himself German at a more accelerated pace. He also read the latest medical papers, particularly on the relationship between cowpox and smallpox.

Brown's Appointment as Naturalist on H.M.S. *Investigator*

Later in 1800, Brown's medical responsibilities increased, so much so that he had little time for botany. It now looked like his destiny was to be a military medical officer but fate again interceded. Originally, the *Lady Nelson* was to be sent to continue the exploration of the interior of New Holland but Flinders reported to Banks that it was still imperative to explore the coasts of New Holland, i.e., to determine if it was one or several islands.

Although the botanical collector, George Caley, was exploring and collecting in New South Wales, Banks found Caley's work to be rather limited. In spite of Banks's misgivings, Caley proved to be a very able botanist in the long run although Banks was not an enthusiastic supporter. Caley became best known as Banks's eccentric collector, and somewhat unfairly as the man who led a failed exploration of the Australian interior. Because he came from a humble background and enjoyed no advocate, he had to act as his own promoter. He always freely expressed his opinions about his abilities, and his frankness or directness ruffled some feathers. When he became bored and dissatisfied with his position as gardener at Kew Gardens and other places in southern England such as Curtis's garden in Brompton, he felt that he must write to Banks and request to be sent to Australia to collect plants. After Park refused to go as naturalist on the *Investigator*, it seemed to Caley that an opportunity had opened up but Banks did not think that Caley was suitable for the position. Eventually, he allowed Caley to travel to New South Wales on the *Porpoise* on a limited salary of 15 shillings a week although the government also did not want anyone with Caley's limited background. He would serve as Banks's collector and would also collect seeds for Kew Gardens. This occurred after Caley wrote a rather abusive letter to Banks, complaining that Banks had caused him injury. Banks did not react, showing some magnanimity and patience and tried to mollify him by eventually trying to do his best for Caley in spite of the sharp words that had been exchanged.[20]

It seemed that the government was ambivalent in supporting such a large undertaking. What finally clinched the argument in favor of funding such an extensive journey under the command of Captain Flinders was the report that a French expedition had already set with a large crew and group of scientists. That survey, sponsored by Napoleon and led by Nicolas Baudin (1754–1803), included 24 scientists exploring the same waters during the same period. Nothing seemed to impress the English as much as the idea that the French were poised to make great strides in exploration in natural history, an expedition that had strong military implications as well. Banks was convinced that an English expedition would be desirable and the Admiralty agreed. Therefore, Banks revived the plan of several years before for an expedition that would explore this area and study its natural history. The goal,

---

[20] See Joan Webb's *George Caley, Nineteenth Century Naturalist: A Biography* (Chipping Norton: S. Beatty & Sons, 1995), for an excellent and comprehensive account of this affair, particularly chapter one, "The Making of a Botanist."

however, would be a bit different than the earlier plans that involved exploration of the continent's interior with Mungo Park. Exploration of the interior would be placed in the hands of lesser figures like Caley, but Flinders convinced Banks that conducting a circumnavigation of the continent together with an ambitious program of collecting flora, fauna, and minerals would be desirable now. And Banks knew just the man who, by virtue of his great ability as an observer and a diligent collector, could make this survey a success, Robert Brown, medical officer in the Fifeshire Fencibles.

Banks thus wrote to Brown on December 12, 1800, informing him that a ship was to be prepared for exploration of the natural history of New Holland, and it was determined "that a naturalist and a Botanic Painter shall be sent in her." Banks informed Brown that if he accepted the appointment, he would recommend him but if he did, "it would be necessary for you to Come here as speedily as you Can." He also informed Brown that his salary would be £400 per year and the voyage would last at least 3 years. He promised Brown he would not recommend anyone else until he heard from him.[21]

Several days later (December 17, 1800), Brown quickly accepted Banks's offer as naturalist, "I lose not a moment .... The situation of Naturalist to the New Holland expedition I will most cheerfully accept," providing he could be given extended leave from his military obligations.[22] Because of Banks's influence, this initially proved to be not much of an obstacle; the only problem was in finding a suitable replacement for Brown. Brown was given his extended leave or so it seemed at the time, assuring his future success, a salutary development for British botany and natural history.

Brown arranged with some of his friends to store his books and on his 27th birthday (December 21, 1800), he set out with his good friend, George Kerr. They sailed for England, and took a coach to London, reaching there on Christmas day, 1800. He was able to dine with Banks the following day and secured lodgings in Rupert Street. He made arrangements to establish a herbarium that would consist of plants from New Holland, drawn from different collectors and from Banks's own collection, and which he would add to considerably from his own collection gathered during his exploration. He had the assistance of Jonas Dryander in completing this task. At night, he copied Daniel Solander's descriptions of plants from New Holland. This was a large project for the list consisted of 620 pages with 620 separate species.

On January 20, 1801, Brown met and dined with Flinders and Ferdinand Bauer who would serve as draftsman on the expedition. It was his first meeting with the men who would be so much a part of his life the next few years. Flinders was promoted to Commander on February 16th and H.M.S. *Investigator*, a sloop weighing only 334 tons, was ready for the sea by March 27th. Flinders's orders were to survey the northern and southern coasts of New Holland, to determine if it was one

---

[21] Banks's letter to Brown, December 12, 1800, *British Museum Add. MS.* 32439.24. Also in *The Indian & Pacific Correspondence*, vol. 5, letter # 161, p. 223.

[22] *Brown Correspondence, British Museum (Natural History, Botany Department)* 3.101. A copy is in *The Indian & Pacific Correspondence*, vol. 5, letter # 163, p. 225.

continent or several islands. He was to first go to King George Sound, then the northwest coasts, the Torres Strait, and then the east coast. Flinders had previously sailed with William Bligh (Captain Bligh), after the unfortunate experience of the mutiny that was the source of Bligh's notoriety. However, Bligh eventually was able to surmount his difficulties and became Governor of New South Wales.

Flinders's earlier experience with Bligh proved to be invaluable. In addition to the experience he gained, he understood the important role natural history would play in this voyage as demonstrated by the fact that he had a prefabricated plant greenhouse constructed on the *Investigator* to keep the specimens that were collected together. This greenhouse would also serve as a herbarium, which Brown started initially from a sampling of Australian plants from Banks's collection.

Banks helped furnish a library for the ship for the officers' use, and included such works as George-Louis Leclerc, Comte de Buffon's *Discours Académiques*, William Enfield's *Institutes of Natural Philosophy*, Flinders's *Observations on the Coasts of Van Dieman's Land*, and Withering's *Outline of Mineralogy*. The *Investigator* was deemed fit and it sailed to Spithead on June 11th. Brown had been working at Banks's library at Soho Square on Solander's list of New Holland plants and meeting with other botanists concerning his earlier work.

Then there was a last-minute complication. Brown was expected to resign his commission; he would not be granted extended leave unless he did so. Previously, there were other cases in which officers had not resigned but were given extended leave. Brown's commander, Colonel James Durham, felt it to be duplicitous behavior on Brown's part. Banks intervened, Brown's extended leave was granted, and he was able to retain his commission as well. Peter Good, foreman at Kew, was hired to assist with the collecting, and a 19-year-old landscape painter and naturalist who had impressed Banks, William Westall (1781–1850), joined the party. John Allen, a miner with whose family Banks was acquainted, was given the post of miner and the responsibility of taking samples of all rocks from mineral veins and bringing them home.

Brown was appointed to be the geologist and zoologist in addition to serving as botanist. Banks wrote to Brown on June 15, 1801, shortly before departure, that he had too good an opinion of Brown's abilities to offer any instructions but merely indicated that it would be best for him to restrict his attention to those branches of natural history "which you are best acquainted with" and devote "full attention to Botany Entemology Ornothology [sic] &c." He also indicated that Caley could join the expedition at Sydney providing that the collections he made be sent to Kew and not be sold. He briefly discussed the other naturalists, in mineralogy for example. The talented artist (draftsman), Ferdinand Bauer, completed the scientific party. There was a crew of 88 chosen by Flinders, and a surgeon, assistant surgeon, and Flinders's cousin, John Franklin, who was a midshipman at the tender age of 15. Brown's duties did not extend to serving as a medical officer, although according to a tradition in the British Navy often observed, the surgeon automatically was regarded as the chief naturalist, but the custom was not followed in this instance.[23]

---

[23] Banks's letter of June 15,1801, to Brown, *British Museum Add. M.S.* 32439.41. Also in *The Indian & Pacific Correspondence*, vol. 5, letter # 288, pp. 363–364.

Brown's diary entry for June 14th 1801 noted, "In the Afternoon set out from London to join His Majesty's Sloop Investigator lying at Spithead." For June 15th he wrote, "Joind the Investigator. Arrivd at Portsmouth about 8 oClock AM. Got on board the Investigator about 10 oClock."[24] This was his first opportunity to see the *Investigator*, which would be home for the next few years. It was not an altogether pleasant sight because the ship was old and not in the best of shape. There was some consolation for Brown because the ship had a good library that he made liberal use of while waiting for the ship to sail, trying to utilize the lag time to his best advantage.

The next month, Brown spent his time at Spithead waiting for the ship to sail. He began reading John Hawkesworth's *An Account of the voyages undertaken by the order of His present Majesty for making discoveries in the Southern hemisphere*, a vast narrative of Captain Cook's first voyage on the *Endeavour*. His entry for the 15th also noted that he remained on the *Investigator*, "occasionally going on Shore to Portsmouth & once visiting the Isle of Wight ... [and] observed a few plants w[hi]ch I have noted in my small octavo Botanical common place Book."[25] He prepared himself for the voyage by continuing to read about "Cook's First Voyage" as well as doing local collecting. Initially, he did not make entries in his diary everyday but soon he began reporting on mostly mundane events. The July 4, 1801, entry does stand out because Brown recorded the details of his excursion to the Isle of Wight where he observed a number of specimens in the orchid family (Orchidaceae), e.g., *Ophrys apifera* and *Ophrys pyramidalis* "in abundance" and *Iris foetida*. He also had a 3 o'clock afternoon dinner with a pint of Port and supper at Nieuport (Newport), the capital of the Isle of Wight, with several glasses of punch. The latter activity was a pattern Brown would repeat in the following days prior to departure. The following day (July 5th), upon his return to the ship, he noted fructification in the lichen, *Diploicia canescens*, in the garden of a nearby inn at Brixton, and *Typha latifolia* (bulrush or cattail) on shore, probably near the ship.[26]

Several days later (July 10th), Brown looked over the plants collected by Johann Forster (the ship's naturalist on the Cook's Second Voyage on H.M.S. *Resolution*), and sent to Banks from Port Jackson, Van Diemen's Land, and Australia's west coast that Brown was allowed to keep in his cabin as "study sets."[27] In addition to reading about Cook's First Voyage and drinking his "usual quantity of wine" and "a glass of grog in the evening," he looked over a few articles in Jean Baptiste de Lamarck's *Dictionaire Botanique*.[28] Brown also reported reading Comte de Buffon's *Théorie de la Terre* and Jean Claude de Métherie's (1743–1817) *Théorie de la*

---

[24] *Diary of Robert Brown*, British Museum (Natural History, Botany Library). The *Diary* has been transcribed completely with considerable annotations in *Nature's Investigator; the Diary of Robert Brown in Australia, 1801–1805*, compiled by T.G. Vallence, D.T. Moore, and E.W. Groves (Canberra, Australia: Australian Biological Research Study, 2001), p. 25. Notes to the *Diary* will indicate the page numbers in *Nature's Investigator*.

[25] *Brown Diary*, p. 25.

[26] *Brown Diary*, pp. 28–29.

[27] *Brown Diary*, pp. 26–27 and 35.

[28] *Brown Diary*, p. 35.

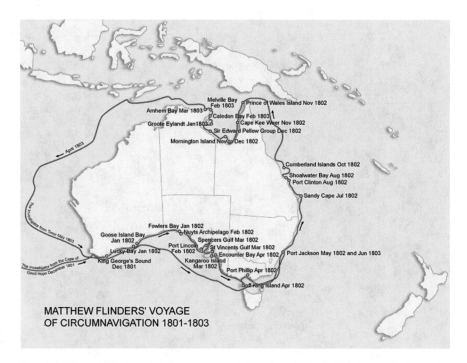

MATTHEW FLINDERS' VOYAGE
OF CIRCUMNAVIGATION 1801-1803

Fig. 5.4 Map of Matthew Flinders Voyage of Circumnavigation 1801–1803, © Centre for Australian National Biodiversity Research

Terre.[29] Lamarck's work may have had a particular impact on Brown in that it reinforced the shortcomings of Linnaeus's artificial system of plant classification he began to develop (Fig. 5.4).

The ship was ready to sail. Brown, in his *Diary*, reported that he went on shore to Fort Monckton—his last extended stay on shore before departure (July 16th)—with Bauer, Westall, and John Crosley, the astronomer for the expedition.[30] The next day, on July 17th, they received their sailing orders, and Brown wrote to his mother, Charles Francis Greville (1746–1809), a politician, patron, and a plant collector and skilled amateur botanist, and Robert Scott (1757–1808), who was Professor of Botany at Dublin from 1785 to 1808.[31] The expedition finally was set to begin.

---

[29] *Brown Diary*, pp. 37 and 40.

[30] *Brown Diary*, p. 40.

[31] *Brown Diary*, p. 41.

Chapter 6
So Remote a Country as New Holland

Contents

Brown noted on June 14, 1801, in his diary, "In the Afternoon set out from London to join His Majesty's Sloop Investigator lying at Spithead."[1] After months of preparation, this low-keyed entry of Brown's marked the start of an adventure that would so profoundly shape his life. He noted in his diary that during the next few weeks, June 14 to July 18, 1801, he continued reading about Cook's first voyage as well as Solander's descriptions of "New Holland plants" and several other works. He recorded drinking a pint of wine, Port, and "cherry" [sherry] at three with dinner along with taking his usual glass of grog during his dinner meal. Finally on July 18, 1801, the *Investigator* departed from Spithead.

Before they made landfall after several weeks at sea with "Madeira in sight," Brown observed (in his diary entry for July 31st), "a small Turtle caught asleep on the surfaces," noting also that it did not seem to belong to any known species and was less than ten pounds in weight. Brown dissected the turtle and found that its circulatory system "corresponded" with the information found in Alexander Munro's *The Structure and Physiology of Fishes Explained and Compared with*

---

[1] *Diary of Robert Brown*, Natural History Museum (Botany Library); also *Nature's Investigator; the diary of Robert Brown in Australia, 1801–1805*, p. 25.

© The Author(s), under exclusive license to Springer Nature Switzerland AG 2021
J. Schwartz, *Robert Brown and Mungo Park*, Memoirs of The New York
Botanical Garden 122, https://doi.org/10.1007/978-3-030-74859-3_6

Those of Man and Other Animals (Edinburgh, 1785).[2] This otherwise unremarkable episode illustrates the range of Brown's interests and knowledge, qualities that would serve him well further along in his journey when he encountered many living forms, most particularly unusual animals very few previously had been privileged to observe.

Madeira

The ship reached Madeira on Sunday, August 2nd, 1801, and Brown, Bauer, and Flinders went ashore at Bujió, the southernmost of the three uninhabited Desertas Islands. The Desertas together with the islands of Madeira and Porto Santo make up the islands in the archipelago collectively referred to as Madeira. Brown noted the very porous quality of the rocks and bioluminescent medusa. The following day (August 3rd), they anchored at Funchal (Funchal Roads), and on the 4th, Peter Good went ashore with Brown, Bauer, and John Allen to see the town as well as explore the interior. They collected many plants in the vicinity of the town (in the vineyards and to the east of the town). Good observed that Madeira was well known, "both [in] its natural History and productions."[3] Afterwards, they enjoyed dinner at "the English Hotel."[4]

Brown, Allen, and Good continued their exploration of Madeira the next few days, including the highest point of the island, Pico Ruivo, which involved a good deal of hiking because it was surrounded by almost equally high mountains. On midday of the 7th, they set sail for the southern tip of Africa. Even though Brown expressed disappointment that he did not find additional plant species from those collected by Banks years earlier, Brown prepared a local list of the flora. Madeira did not have the same impact on Brown that it had on Huxley when he served as surgeon and naturalist on H.M.S. *Rattlesnake* years later (1846). Huxley noted in his *Diary* after having his first taste of the tropics when he was also voyaging to Australia: "A fair wind is carrying us fast away from Madeira .... The feelings produced by the rival beauties of its scenery would require for their fit expression more eloquence than I possess .... Mountain scenery is new to me, and the semi-tropical

---

[2] Alexander Munro (1733–1817) was another product of Edinburgh University and its medical school. He was a member of the Harveian Society and joint secretary of the Philosophical Society along with David Hume and became its secretary when it became the Royal Society of Edinburgh. After Munro's training in Edinburgh, he studied in London with William Hunter, and in Berlin and Leiden as well. The reference to catching the turtle is in *Robert Brown Diary*, p. 54, and in the *Journal of Peter Good*, where Good reported that it was actually Captain Flinders who caught the turtle, p. 36.

[3] Peter Good. *Journal of Peter Good: Gardener on Matthew Flinders Voyage to Terra Australis 1801–1803* Phyllis I. Edwards ed. (London: British Natural History Museum, 1981), p. 38.

[4] *Diary*, p. 58.

vegetation, the banana, the cactuses and the palm trees, call to mind all that I have ever read of foreign countries and give me a foretaste of the far south.[5] However, Brown was not similarity moved by Madeira's "rival beauties."

Cape Town

After leaving Madeira, the *Investigator* set sail toward the Cape of Good Hope (Simon's Bay).[6] They crossed the equator and Flinders chose to observe the usual custom of "ducking and shaving" for members of the crew who were crossing the equator for the first time. Brown took a jaundiced view of the proceedings perhaps similar to Huxley's feelings about what he felt was such a childish custom 40 years later when he sailed on the *Rattlesnake*. Brown commented in his diary in his entry for Tuesday, September 8th, "Crossed the line in about 17 [°] W[est] Longitude the usual ceremonies."[7] Good also reported that the crew "performed the usual Ceremony of Shaving for having Crossed the Equator and as usual also the Sailors got drunk and turbulent at night."[8] Brown ignored these activities although it is not clear from his account whether he was merely a spectator or actually suffered some of the indignities of this initiation rite. He busied himself by studying some of the New Holland plants he had taken with him. There were further references to some of the birds encountered as they sailed south and Peter Good's French lessons were mentioned. There was the suggestion that Brown was doing the tutoring.

Land was sighted on October 16th and they anchored at Simon's Bay at the Cape of Good Hope (at a region called "False Bay"). Brown went ashore the next day and noted the "great number of plants," forms he had never seen before, and noted the "number & beauty of heaths striking."[9] Brown's mood was lightened by the plants he observed, "for [their] variety and beauty were beyond description."[10] Brown, Bauer, and Allen collected insects and minerals in addition to the varieties of plants they assembled. The naturalists collected a wide variety of plants—"Orchis, Drosera, and Hemimeris" for example. For the remainder of October, Brown remained on board working on the specimens he gathered while Good collected additional specimens for him to examine. In other instances, Brown walked in the area, and once he traveled from Tokai, a Dutch farm, to Paulsberg, and then to Cape

---

[5] *T.H. Huxley's Diary of the Voyage of H.M.S. Rattlesnake*, ed. by Julian Huxley (London: Chatto and Windus, 1935), pp. 17–18.

[6] After the entry for August 5th, he provided only summaries for the remainder of the month, until September 3rd, which is on the top of the next page. See *Nature's Investigator*, p. 61.

[7] Brown's attitude was somewhat reserved concerning this ritual although there he acknowledged Flinders's claim that it was "part of his plan for preserving the health of his people," entry for Monday, 7 September 1801, p. 62.

[8] *Journal of Peter Good*, p. 40.

[9] *Diary*, p. 78.

[10] *Journal of Peter Good*, p. 42.

Town over Table Mountain. Good reported shooting "several beautiful birds and killed a large dangerous serpent, possibly a Cape Cobra."[11] Although this work proved rewarding it was quite fatiguing. They did a great deal of hiking on barren sands and in the nearby mountainous terrain. Brown took a nasty fall and hurt his leg but apparently there was little lasting damage as he was eager to take advantage of the great variety of riches in natural history set before him.[12] The initial impact the strikingly beautiful and unusual surroundings had on these European naturalists is apparent by reading accounts of their first exposure to lands so different from what they were accustomed to seeing.

During the last week at the southern tip of Africa, Brown and the others explored the mountain range (Steinberg) in the Cape Town vicinity, visiting Table Mountain and Devil's Mountain, regions rich in plant life, and Brown and Good eagerly collected samples of it. Their ascent up Table Mountain was made more difficult by the persistent rain and fog. They found several varieties of heaths, various species of shrubs, different species of ferns, e.g., *Polypodium capensis* (now called *Cyathea capensis*), and several mosses belonging to the genus *Orthotrichum* Hedw. (*Orthotricha nonnulla*) at lower elevations. Shortly before departure, Good went ashore and found a species of orchid that had not been seen before, and he and Brown found some sample of "granite plants" (possibly *Eucalyptus obliqua*, growing in poor soil near granite). Brown also encountered two freshly discovered species belonging to the Proteaceae family, the tropical and subtropical evergreen flowering shrubs and trees, notable for its nuts (like macadamia) that were included in the genus *Banksia*, *Serruria foeniculacea*, and *S. flagellaris*. Brown was pleased with the work he accomplished so far and eagerly looked forward to their arrival in New Holland. The *Investigator* set sail on November 4, 1801, from False Bay, sailing across the Indian Ocean. A decade earlier, Captain George Vancouver (1758–1798) had sailed on the *Discovery* along the southern shore of Australia. Vancouver's surgeon and naturalist, Archibald Menzies (1754–1842)—a Scot who had studied botany and medicine at Edinburgh University and a friend of Brown as well—had made the first collection of plants from the region.[13]

---

[11] *Journal of Peter Good*, p. 43.

[12] For a lengthy account of Brown's stay at Simon's Bay in addition to Brown's diary and Good's journal, see John P. Rourke's "Robert Brown at the Cape of Good Hope," *Journal of South African Botany* 40 (1974): 47–60.

[13] David J. Mabberley indicated that there had never been more "systematic" and "sophisticated" collecting as exhibited on this voyage, and it can be assumed that much of it was due to Brown's meticulous planning and preparation, *Jupiter Botanicus*, p. 81. Earlier travels, particularly Cook's voyages in this region, served as a foundation for Banks's herbarium and was one of the motivating forces for the expedition on the *Investigator*.

Western Australia

Brown collected more samples of Proteaceae from islands in the Indian Ocean while the leaky ship made its way toward Australia. There was a minor delay that potentially could have been worse when the ship's cables became bent, and the expedition was faced with the possibility of having to land and spend time repairing the cables (at the St. Paul-Amsterdam Islands in the southern Indian Ocean). However, they were able to postpone solving the problem until later on. Brown did not make any note of this episode; his diary entries were quite sparse during the month of November before they made landfall in Australia. They anchored in the southwestern corner of western Australia on December 8th (1801), at King George Sound near Seal Island at Point Possession (on the south side of the entrance to Princess Royal Harbour at the southwestern tip of Australia). Brown was struck by the diversity of the vegetation and the unparalleled floral display. He was not the first European naturalist to visit the shores of southwestern Australia. The Dutch explored this territory in the seventeenth century, and both Menzies and Vancouver (previously cited) were there a decade earlier. The French also arrived prior to the *Investigator's* making landfall there in late 1801. But Flinders's visit had the most impact in view of the advances made in natural history, largely because of the efforts of Brown and his naturalist-colleagues on the voyage.

The following day (December 9th), Brown, Good, and others disembarked after breakfast and explored the south shore, collecting specimens, i.e., samples of seeds and geological samples with fossils of "univalve shells" (gastropods) in the limestone soil. The King George region, situated on the Indian Ocean, proved to be the most abundant part of Australia botanically, consisting of 87% of the species and 41% of the genera Brown collected on the continent.[14] After allowing Good to go ashore the following day, on the 11th, Brown reported "more successful botanizing."[15] Beginning at Seal Island—described by Flinders as a mass of granite accessible only at its western end—Brown was able to collect approximately 500 species during the height of the Australian summer during the three weeks they were there.[16] In his written account of the botany of the voyage, Brown noted in an appendix to Flinders's *A Voyage to Terra Australis*, "Our first anchorage was in King George Third's Sound .... we remained for three weeks, in the most favourable season for our pursuits; and our collection of plants made chiefly on its shores and a few miles

---

[14] See Nancy Tyson Burbidge, "The Phytogeography of the Australian Region," *Australian Journal of Botany* 8 (1960): 75–211.

[15] *Diary*, p. 93.

[16] Matthew Flinders. *A Voyage to Terra Australis: Matthew Flinders' Great Adventure in the Circumnavigation of Australia*, Tim Flannery ed. (Melbourne, Australia: The Text Publishing Company, 2000). vol. I, p. 54. Flinders noted, "wood seemed not to be abundant near the shore." *A Voyage to Terra Australis* originally was published in London in 1814, the year Flinders died (London: G. & W. Nicol. 1814).

into the interior of the country, amounts to nearly 500 species, exclusive of those belonging to those of the class Cryptogamia."[17]

A fire made by aborigines near Princess Royal Harbour attracted the ship's party enabling them to have their first contact with the native people (on December 14th). Brown noted their language, and observed their customs even though this initial contact seemed awkward and even menacing because the native Brown came in contact with became startled. The man brandished his spear and then ran quickly into the bushes near the beach. He proceeded to set fire to the bushes so he would not be followed and he ran away with what was thought to be his family. The following day, there were further contacts and the natives exchanged a spear and a stone hatchet for some of the trinkets Flinders's crew had available. It seemed that they were not much interested in what the Westerners had to offer, with the exception of nightcaps and handkerchiefs, and they favored the ones that were red in color. Westall was able to make his first sketches of the aborigines they encountered.

In the middle of the month (December 15th–17th), the crew stocked up on many oysters they collected at the aptly named Oyster Harbour, and Brown described some of the fish he observed while remaining on board, particularly a specimen named Brown's leatherjacket (*Acanthaluteres brownii*).[18] On the 17th, Brown, Allen, and Good hiked along the shore of Oyster Harbour in search of the river described by Vancouver a decade earlier. The going was difficult because they found themselves in mud in an extensive salt marsh but they were able to observe many wild ducks and a kangaroo in their labored efforts. Brown noted some of the largest trees he had ever yet observed, *Eucalyptus resinifera*.

Before Christmas, they went on an extensive exploring and collecting trip to Torbay Inlet. The last leg on the 24th was quite exhausting and they ran low on water because the sun was quite hot. Brown and Good were quite parched and Bauer found himself unable to go any further. This acted as an impediment in conducting their explorations. They had to sleep on the beach on Christmas Eve, and arrived back on the ship on Christmas day where they enjoyed their Christmas dinner. The ship was being refitted for the next part of the trip to the Recherche Archipelago, a spectacular seascape formed by volcanic activity and known for its biodiversity. In the meantime, the collection of plants and seeds native to the area continued unabated, and Good went ashore to collect specimens while Brown remained on the ship cataloguing the specimens that had been already collected. Good eventually sent numerous samples of seeds he collected to Banks.

---

[17] Cryptogamia is a subdivision of the plant kingdom containing plants that do not bear seeds, i.e., ferns, mosses, algae, fungi, and lichens.

[18] A fine illustration of the fish is shown in Sir John Richardson's *Ichthyology of the Voyage of H.M.S. Erebus and Terror* (London: Longman, Brown, Green, and Longmans, 1846). Richardson (1787–1865) was another Scottish naturalist-physician. He was a famous Arctic explorer, and was very instrumental in helping Thomas Henry Huxley get a position as surgeon and naturalist on H.M.S. *Rattlesnake* in 1845–1846.

Prior to their departure from Torbay Inlet, Brown did some exploring of the southern part of Princess Royal Harbour, Vancouver Peninsula, particularly "a little island" described by Cook on his first voyage.[19] Brown and Good spent the time packing the specimens they collected, including the brilliant-colored *Banksia repens*, a member of the Proteaceae family. Brown also conducted some ethnological studies, observing the natives when they permitted him to do so, and he listed some of the words from their language in his diary. Brown's diary entry for January 2nd reveals that he remained on board most of the time, describing the plants collected, and reported that "Mr. Good went in search of the pitcher plant w[hi]^ch Messrs Bauer & Westall had found yesterday in flower," adding "he returned with it in the evening."[20]

Southern Australia

On January 4th they were ready to sail again, and Brown left a bottle with a note summarizing the progress made so far on Seal Island, exploring it one last time in the afternoon. He also described a number of plants he had not seen before, including a mallow belonging to the Malvaceae, *Lavatera mollissima* (actually *L. plebeia* Sims); a type of peppergrass or pepperweed belonging to the mustard family, Brassicaceae, *Lepidium dentatum*; and a member of a group of medicinal herbs, *Lobelia insularis* or *L. alata* Labill. in the Campanulaceae. Flinders noted, "Amongst the plants collected by Mr. Brown and his associates, was a small one of a novel kind, which we commonly call the pitcher plant."[21] The following day, the *Investigator* sailed eastward toward Lucky Bay, east of Esperance Bay. On the way, they stopped frequently at inlets and small islands, such as "The Doubtful Islands" and Mount Barren, and on the 8th of January, they reached the Recherche Archipelago (Figs. 6.1 and 6.2).

Brown collected over a hundred new species including a feather flower appropriately named *Verticordia brownii*, commonly known as pink or wild cauliflower (a member of the Myrtaceae, the myrtle family). They spent January exploring the islands in the Recherche Archipelago, including Middle and Goose islands. After the abundant plant life Brown found in King George Sound and Seal Island, this area proved to be a bit of a disappointment though Brown's total of new plant species collected now reached approximately 700. It was an area of rugged beauty but the volcanic soil may have inhibited the growth and variety of plants.

Brown reported that many became ill eating the nuts (or seeds) of the cycad, *Zamia*, but he personally did not suffer any ill effects. Both Brown and Good made the most of their visit there, observing and collecting on the shores of Lucky Bay,

---

[19] Geak Point on Vancouver Peninsula. Entry in *Diary* for December 28th, p. 103.

[20] *Robert Brown Diary*, Botany Department, British Museum (Natural History), p. 106 in *Nature's Investigator; the Diary of Robert Brown in Australia, 1801–1805*.

[21] It was possibly a species belonging to the genus *Drosera*, referred to by Darwin as "his beloved" plant later on. Flinders, *A Voyage to Terra Australis*, I (London: G. & W. Nicol, 1814), p. 64.

Fig. 6.1 *Verticordia brownii*, watercolor by Adam Forster, National Picture Library of Australia

Thistle Cove, and hiked-up Frenchman Peak. Good and Brown found several species of *Banksia* near Lucky Bay, including *Banksia speciosa* (Handsome or Showy Banksia). Flinders recorded that "the botanists landed in the morning [of January 14, 1802] upon Middles Island for I had determined to stop a day or two as well for their accommodation as to improve my chart of the [Recherche] archipelago." He added that the northern island was "covered with tufts of wiry grass intermixed with a few shrubs. Some of the little blue pinguins, like those of Bass Strait, harboured under the bushes …." (Figs. 6.3 and 6.4)[22]

The Great Australian Bight

They left Recherche Archipelago in the middle of January and entered the Great Australian Bight, on January 17, 1802—a region named by Flinders which was characterized by a large open bay about 720 miles long, extending from Western to Southern Australia, consisting of large cliffs: "white sand hills succeeded by

---

[22] Flinders, *A Voyage to Terra Australis*, I, p. 87.

Verticordia brownii

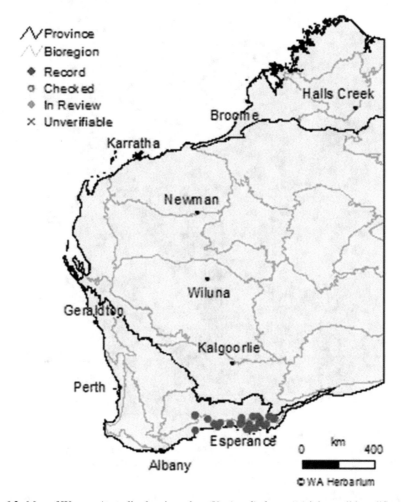

Fig. 6.2 Map of Western Australia showing where *Verticordia brownii* (pink or wild cauliflower) is still found, and the region where Brown collected it in January 1802. Image used with the permission of the Western Australian Herbarium, Department of Biodiversity, Conservation and Attractions (https://florabase.dpaw.wa.gov.au/help/copyright. Accessed on Monday, 27 July 2020)

perpendicular cliffs of whitish stratified stone."[23] The *Investigator* continued its eastward course, sailing along the Bight to its head, and eventually anchored at Fowler's Bay at the end of January (28th).

Although the Dutch originally discovered the region in the seventeenth century, Flinders made the first detailed charts of the area and named it for his First Lieutenant, Robert Fowler. Brown had made very brief and generally

---

[23] *Brown's Diary*, p. 119.

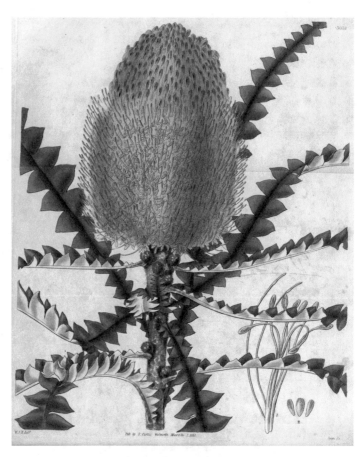

Fig. 6.3 *Banksia speciosa* (Handsome or Showy Banksia) in Proteaceae family from *Curtis Botanical Magazine*, described by William Jackson Hooker, Regius Professor of the University of Glasgow (1831), New York Botanical Gardens, The LuEsther T Mertz Library

uninformative entries in his diary during this part of the journey but once the *Investigator* arrived at Fowler's Bay, Brown went ashore and described the geology of the place in some detail most notably, formations containing dark granite which resembled basalt, and unusually colored species of blue and black ants tinged with red. He found few new species of plants but observed a great quantity of seagrass, likely *Zostera muelleri*.

The ship then sailed on to St. Francis Island of Nuyts Archipelago (February 2nd), situated on the eastern part of the Great Australian Bight, where they observed many birds, especially petrel. The weather was unbearably hot—it may have exceeded 125 °F (approximately 52 °C)—so exploration was limited to the southern regions. They did observe numerous species of birds, and other wildlife such as kangaroos. Brown continued gathering up specimens of animals as well as plants, particularly birds like the New Holland honeyeater (*Phylidonyris novaehollandiae*), a bird native to southern Australia, 17–18.5 cm long and mainly black and white

Fig. 6.4 *Banksia speciosa* (Showy Banksia) in Proteaceae family by Sarah Anne Drake from *Edwards Botanical Register* New Series, Volume VII (1835), New York Botanical Gardens, The LuEsther T Mertz Library

with a large yellow wing patch and yellow sides on the tail, "several birds of the petrel kind," incorrectly labeled by Brown as *Procellaria norfolkiensis* (Flinders referred to it as a "sooty petrel"), and the usual sample of kangaroos.[24] Both Brown

---

[24] Flinders, *A Voyage to Terra Australis*, I, p. 108, *Journal of Peter Good*, p. 57, and *Nature's Investigator*, p. 125. See the editors' notes about Brown and Flinders's incorrect scientific attribution.

and Good reported (February 6th–7th) that they anchored at the largest island in Nuyts Archipelago (St. Peter Island, 2 miles long or 4–5 leagues in length), and "saw a great number of quadrupeds of the kangaroo kind" but "no variety of plants."[25] Flinders took advantage of this relatively unexplored territory to name various landmarks after members of the crew: "There remained forty miles of space between Point Bell and Point Brown. West of Cape Bauer ... there is a low island ... which I called *Olive's Island*. Another cliffy, and somewhat higher projection ... I named *Point Westall*."[26]

Brown continued his observations of animal life of southern Australia, and was particularly drawn by the behavior of seals in Waldegrave Island off the Eyre Peninsula (February 11th–12th), and the territorialism of the male seals. The males were aggressive toward members of the party including Brown, and some of them were unfortunately killed: "At first were oblig'd to defend ourselves from the males but afterwards I confess wantonly attackd & even killd several of them."[27]

They stopped at "a very large island" on the 12th and Flinders named it Flinders Island—part of the Nuyts Archipelago in the "*Investigator Group*"—after his brother Samuel Ward Flinders, who was Second Lieutenant on the *Investigator*. They continued their eastward journey toward Spencer Gulf and the Port Augusta area, finding the terrain of certain parts of the shore of Eyre Peninsula, Coffin Bay, and Whidbey Islands difficult to walk because of the granite and consequently it was "quite sterile." Bushes or shrubs, probably a variety of *Eucalyptus*, covered much of the shoreline. In addition, Brown discovered a new species of *Eucalyptus* in the Neptune Isles. Flinders reported finding a species of kangaroo whose members were quite small, "not bigger than a cat, [and] was rather numerous. I shot five of them, and some others were killed by the botanists and their attendants ...."[28]

The *Investigator* entered Spencer Gulf on February 20th, and on a wooded island Brown found new species of *Eucalyptus*, mangroves, and thickly tufted grass but no freshwater. On February 22nd Flinders began to search in vain for several of the crew who embarked on a small craft, a "red cutter." Included in this party was John Thistle, the ship's Master, Midshipman, and master's mate, William Taylor, and six other seamen on board which went across Thorny Passage to search for a suitable place for anchoring the previous day.

Prior to this voyage, Thistle had accompanied Flinders on earlier journeys so his loss was a serious blow for Flinders but this unfortunate event again emphasizes the dangers inherent in conducting expeditions in the late eighteenth century in relatively unexplored territory. Flinders wrote: "From the heights near the extremity of

---

[25] The kangaroos were probably tammar wallabies (*Macropus eugenii*). *Journal of Peter Good*, p. 58, *Robert Brown Diary*, Botany Department, British Museum (Natural History), and *Nature's Investigator*, pp. 126–127.

[26] Entry for February 5th, 1802, Flinders, *A Voyage to Terra Australis*, I, p. 111.

[27] *Nature's Investigator*, p. 130. Today, the seal population in the Eyre Peninsula is threatened by pollution and the impact of the human population, particularly the aquaculture industry.

[28] Flinders, *A Voyage to Terra Australis*, I, p. 125.

Cape Catastrophe, I examined with a glass the islands lying off for any of our people, but in vain." The following morning (on the 23rd), he "went in a boat along the shore …. in one place I picked up a small keg, which had belonged to Mr. Thistle, and also some broken pieces of the boat; but these were all that could be discovered …. I called these *Taylor's Isles*, in memory of the young gentleman who was in the cutter with Mr. Thistle."[29] They spent the last part of February and early March (February 23rd to March 6th) first at Memory Cove, where Brown collected six or seven different plants to add to his collection, and then at Port Lincoln where the Captain left a copper plate "recording the unfortunate loss."[30]

On March 4th, the Captain went ashore with a telescope, possibly to continue his search for Thistle and his party but also he took time to observe an almost total eclipse of the sun, an illustration of Flinders' broad scientific interests. There were no specific instructions indicating how Flinders should proceed in making these astronomical observations. One may assume that Flinders was motivated out of his own scientific curiosity. Flinders conducted this collection of data while he was still grieving over the loss of members of his crew, particularly a man like Thistle who had accompanied him on many voyages since 1794 and was fond of.[31]

Brown remained on board cataloging the specimens he collected in the region while Good went ashore where he reported observing some emus and collected a single specimen of a flowering climber, *Convolvulus erubescens* (Australian bindweed). The *Investigator* continued exploring Spencer Gulf (e.g., Spalding Cove, Kirkby Island) and arrived in the Port Augusta area at the head of Spencer Gulf on March 9th. Good noted the large mountain chain there, "the most considerable we had seen in New Holland."[32] The next day (March 10th), a party of seven including Brown, Good, and Bauer climbed mountains with the highest elevation. This proved to be very arduous because the first part was conducted through mud and seaweed. Flinders noted, "The ascent to Mount Brown had proved to be very difficult …. Mr. Brown found the stone of this ridge of craggy mountains to be arillaceous [having a high clay content] …."[33] Brown recorded an extensive list of the plants he collected during the climb, "Florula Spencer's Gulph, March 10–11, 1802."[34]

---

[29] Flinders, *A Voyage to Terra Australis*, I, p. 137 and *Journal of Peter Good*, p. 64. The cove was renamed "Memory Cove" and Flinders left a copper plate describing the details of the misfortune, at least as much as he knew. Brown listed the names of the crew members lost, in his diary, *Nature's Investigator*, p. 140.

[30] Entries for February 23rd, 1802, Flinders, *A Voyage to Terra Australis*, I, p. 137, and 24th, *Nature's Investigator*, p. 144.

[31] See David J. Mabberley, *Jupiter Botanicus,* where he notes that the solar eclipse "was dwarfed by the death of John Thistle," p. 89. Flinders in *A Voyage to Terra Australis*, included the data of the eclipse but it is just recorded without any additional comment, I, p. 145.

[32] *Journal of Peter Good*, p. 65.

[33] Flinders, *A Voyage to Terra Australis*, I, p. 159. Flinders named this peak for Brown. It still is known as Mt. Brown, and makes up a part of the "Flinders Ranges."

[34] Later on (in 1819), it was inserted in the appropriate part of his diary, *Nature's Investigator*, p. 157.

Encounter with the French

The *Investigator* continued to work its way southeastward, stopping at Kangaroo Island, St. Vincent Gulf, where they spent several weeks (March 21st–April 7th) exploring the island and the rest of this previously unexplored gulf. Flinders noted that the scientific gentlemen landed again to examine the natural productions of the island, as Brown went ashore at the northern end of the island—on Nepean Bay— and observed the vegetation and the numerous birds and seals, noting that the island was composed of micaceous schist and the soil was fairly good. Although the trees were not numerous, he found some new species or varieties of *Eucalyptus* and a small tree uniquely native to southern Australia, the she-oak (*Allocasuarina verticillata*). Also, they observed mimosa, pelicans, brown pigeons, sea lions, and of course numerous kangaroos, probably the Western Grey Kangaroo, which gave Flinders the idea in naming the island. As was the practice then, the birds were often shot with little thought given to attempt to capture them live for more accurate study.[35] Brown sent Good ashore the next day (March 23rd) to continue observations and collect plant specimens, eventually gathering at least eight new species of *Eucalyptus*. Flinders explored the Gulf (St. Vincent) at the end of the month, not only to map this rather undiscovered territory but also to search for sources of freshwater (Fig. 6.5).

Fig. 6.5 View of north side of Kangaroo Island, March 1802, by William Westall, 1811, ©National Maritime Museum, Greenwich, London, ZBA7938

---

[35] Flinders, *A Voyage to Terra Australis*, I, p. 170 and *Nature's Investigator*, pp. 165–166.

On April 4th Flinders reported that he was accompanied by the naturalist [Brown] in a boat expedition to the head of the large eastern core of Nepean Bay.[36] They returned to another part of Kangaroo Island (the eastern part) before they left the Gulf for good on April 7th, arriving at Encounter Bay on April 8th. There, they met the French ship, *Le Géographie*, a ship of the corvette class, part of the French expedition and commanded ably by Nicolas Thomas Baudin (1754–1803).

The store-ship, *Naturaliste*, commanded by Jacques Felix Emmanuel Hamelin (1768–1839), was not in Encounter Bay when Flinders and Baudin met. There was some initial nervousness but in keeping with the spirit of international cooperation in science—particularly because Banks among others tried to foster that spirit of cooperation—Flinders went on board the French ship with Brown acting as interpreter. The interview was conducted mainly in English but Baudin's knowledge of English was not always reliable and similarly Brown's French sometimes failed him.[37] Still they got on well and were able to exchange information the best they could. Baudin indicated that he lost a boat with eight men in Bass Strait (in Victoria), and told of further calamities such as the loss of men at Timor from sickness, particularly his gardener, Anselme Riedlé, whom he was very fond of and was irreplaceable. Baudin indicated that he had Riedlé buried alongside David Nelson who had died in 1789 while serving as the botanical collector on H.M.S. *Discovery* during Cook's third voyage.

Brown observed the meagre collection of specimens, and Flinders noted that the surveying information the French had gathered was quite below standards. Both Baudin and Flinders and their parties got on well, obviously assisted by the fact that the Peace of Amiens had been just signed in March of that year (1802), thereby ending hostilities between their countries. Brown received additional information on a subsequent visit to *Le Géographie* he made the next day (April 9th), concerning discoveries the French made in the Van Diemen's Land area, such as where to find wood and drinking water, and Brown shared information about the wildlife they had encountered so far in the southwestern and southern regions of the continent.

After their cordial meeting, they parted; the French headed west while Flinders headed on an eastern course toward Kings Island in the Bass Strait, and toward New South Wales on the east coast of the continent.[38] They sailed past "The Coorong," which still remains a vast, unspoiled ecosystem made up of freshwater lakes, wetlands, beaches, as well as saltwater.

---

[36] Flinders, *A Voyage to Terra Australis*, I, p. 182.

[37] K.A. Austin noted in *The Voyage of the Investigator 1801–1803, Commander Matthew Flinders, R.N.*, that Flinders could not speak French and for this reason he took Brown, an "accomplished" linguist, with him when meeting with Captain Baudin (Adelaide: Rigby Limited, 1964), p. 108.

[38] Afterwards, both Captains were to share rather sharp reversals of fortune: Baudin died of tuberculosis the following year (1803), and Flinders was captured by the French when hostilities resumed between Britain and France. See Anthony J. Brown's *Ill-starred Captains: Baudin and Flinders* (London: Chatham, 2001).

Bass Strait, Victoria

In the latter part of April, the expedition arrived in Victoria, the south easternmost region of Australia, entering Bass Strait on the 22nd; on the 23rd they anchored a few miles from shore at Kings Island. Flinders, Brown, and Westall went ashore with a cutter and explored the area. They found and killed several small marsupials that Good described as similar to opossums, which Brown identified as "womats," probably the common wombat (*Vombatus ursinus*). The following day (April 24th), Good joined the party as they went further inland and encountered nine different species of ferns. The wooded portion was so dense that it was difficult to penetrate very far but they discovered "a fine lake of fresh water about a mile in circumference," always a fortuitous event on such an expedition.[39] A few days later (April 26th), they were nearly grounded on a shoal near Prince Philip but were able to avoid disaster and they entered deeper waters after a rather close call.

Flinders reported "the narrowness of the entrance did by no means correspond with the width given to it by Mr. Bass," indicating, "it was the information given by Captain Baudin, who coasted along from thence with fine weather, and had found no inlet of any kind" was more helpful in navigating this passage.[40] Flinders also reported that lieutenant John Murray—Captain of the *Lady Nelson*, a vessel that would help convoy the *Investigator* after their departure from Port Jackson later on—had discovered the port 10 weeks before, and had given it the name of Port Phillip (after Arthur Phillip, first governor of New South Wales), and the rocky point on the east side of the entrance, Point Nepean (after Sir Evan Nepean, First Secretary of the Admiralty).[41] The following day, they arrived at Port Philip, and Flinders, Brown, and Westall went by cutter to explore the area, and Good, Bauer, and Allen climbed a steep hill so they could get a better vantage point to survey the land. It was earlier named "Arthur's Seat" because of its resemblance to a hill in Edinburgh, Scotland, of the same name. They found a good deal of freshwater, and the weather was pleasant so their mood improved. They found ample supplies of oysters—probably the mud or flat oyster, *Ostrea angasi*—as well as other species of shellfish. These positive developments, the pleasant weather with more moderate temperatures and sufficient food and water, insured that the ship would be well stocked for the last part of their journey to New South Wales.

Not allowing any distractions despite the fine weather and more comfortable environment, Brown continued his observations on the plants of Victoria unabated, e.g., *Eucalyptus* trees, *Banksia integrifolia* (coastal Banksia) with its beautiful golden-yellow flowers belonging to the Proteaceae, and new species of birds including Brown's slip, actually the glossy black cockatoo and a "new white-crested species," possibly a white cockatoo (referred to as a corella in Australia). They also observed the region's geology, including the finely grained and reddish-gray granite

---

[39] *Journal of Peter Good*, p. 74.
[40] George Bass (1771–1803) explored the area several years before.
[41] Flinders, *A Voyage to Terra Australis*, entry for April 26, 1802, I, pp. 211–212.

and briefly analyzed the quality of the soil, noted by Good in his journal.[42] Brown collected and noted 96 different species of plants belonging to 74 genera in Victoria. Because of his achievements in discovering, collecting, and cataloging plants in this region, Brown to this day has retained the honorific title "Father of Victorian Botany" and there are many places in this territory associated with his name.[43]

Flinders was quite anxious to get to Port Jackson soon because he was concerned about the winter weather and having sufficient food and water: "The time was fast approaching when it would be necessary to proceed to Port Jackson both on account of the winter season, and from the lack of certain provisions."[44] There were some problems with the ship in Port Phillip (the area near present-day Melbourne), so he had additional reason to push on toward Port Jackson (present-day Sydney). In the meantime, Flinders realized that it would be difficult to maneuver the ship in surveying the region so it would be better to perform those tasks from a cutter. This gave Brown additional opportunity to make further observations, and the other naturalists were able to conduct "shore work," (April 28–29th), with Good recording many different species of birds, "an astonishing quantity," including black swans and pelicans.[45] Before they went through the Bass Strait again, Brown went ashore on Mornington Peninsula and Swan Bay (May 2nd), and reported seeing a good amount of *Banksia integrifolia* (coastal Banksia), *Casuarina equisetifolia* (she-oaks), and *Mimosa odoratissima* (often given the genus *Albizia*), much grass and herbaceous plants like geranium.

The following day (May 3rd) they departed Port Phillip and with a strong westerly wind assisting them, they proceeded on their way to Port Jackson in New South Wales. The first part of the expedition had been completed; it was a promising beginning with a well-stocked ship with many specimens, and a successful survey of the southern reaches of the continent. The laggard pace of the French was due to their slow start in getting to Mauritius and a subsequent loss of men from illness. These setbacks enabled Flinders to precede the French in exploring much of the southern coast.

Flinders's single-minded determination to conduct a thorough investigation of this region was critical in allowing his party to discover more than if he had headed

---

[42] *Journal of Peter Good*, pp. 76–79, and *Nature's Investigator*, pp. 190–192.

[43] James H. Willis first bestowed this accolade on Brown in his "Robert Brown's Collectings in Victoria," *Muelleria* 1(1956): 51–54. In this tribute, referred to the southern province of Australia. Helen Hewson's monograph, *Brunonia australis: Robert Brown and his Contributions to the Botany of Victoria* (Canberra: The Centre of Plant Diversity Research, 2002), explained that Brown's contributions to Victoria's botany were accomplished largely by default, because he, unlike others, "divided into very major biogeographic groups," p. 2. Hewson also indicated that "Brown's contributions to the botany of Victoria embraces his taxa from an Australia-population point-of-view, rather than a parochial/regional one giving a more realistic idea of his contribution." Because Brown's "more cosmopolitan/'bigger picture' view ... his delimitation of families and genera" were "of more significance to him than the 'minutiae' of the species/flora of a small region of Australia," p. 2.

[44] Flinders, *A Voyage to Terra Australis*, I, p. 208.

[45] *Nature's* Investigator, p. 193 and Journal *of Peter Good*, p. 76.

for Port Jackson more quickly, and thus his efforts were more successful than the French. His chief object was to survey the northern and southern coasts of the territory and try to ascertain whether New Holland was a continent (a single land mass), or a group of islands intersected by seas from the Gulf of Carpentaria. Ultimately, he was successful in determining that it was a continent. It was not clear to Flinders what the French wished to accomplish from their expedition. In spite of Brown's efforts as a translator, communication was difficult because of their inability to understand each other's language clearly enough.

Port Jackson, New South Wales

The *Investigator* was in sight of Port Jackson on May 6th and they entered the harbor several days later, anchoring in Sydney Cove on Sunday, May 9th. Flinders was relieved that they reached the port before the winter storms because his supply of food and water had run rather low. Flinders noted, "At one o'clock we gained the heads, a pilot came on board, and soon after three the Investigator was anchored in Sydney Cove."[46] At the time, Port Jackson was mainly a penal colony; a third of the inhabitants were convicts. Good noted that the houses were both made of brick or wood, and observed that the houses with their gardens were very nice in appearance despite the town's main function as a penal colony. The region north of Port Jackson was heavily wooded and still controlled by the aboriginal tribes.[47] Brown made few entries in his diary, confining his writing efforts to letters he sent to Banks, Dryander, and Greville. He informed them, in some detail, of his progress. In the harbor at that time, in addition to *Le Naturaliste*, was the whaler, *Speedy*, adding to the dramatic setting when the crew and officers had time to enjoy land in the growing settlement.

Flinders went to see the Governor of New South Wales (Philip Gidley King) almost immediately after disembarking to inform him about the journey so far and his meeting with Baudin's party, which was of course of great interest to the Admiralty. Brown, in the meantime, went to meet Banks's collector, George Caley, and tried to persuade him to join the expedition, but had little success in convincing him. Brown wrote to Banks and informed him of Caley's refusal (at the end of the month, May 30th). Banks realized by then that it would be better for Caley to continue collecting in the Port Jackson area, i.e., it would be more a suitable match for his abilities.

Brown's letter to Banks (May 30, 1802)—the first to the expedition's sponsor since leaving England—informed him about what he and the other naturalists had accomplished, particularly his work with plants which he considered quite successful: i.e., approximately 750 species of plants collected, excluding cryptogams, on the southern coast, supplemented by Bauer's 350 drawings of plants and 100

---

[46] Flinders, *A Voyage to Terra Australis*. I, p. 226.

[47] *Nature's Investigator*, pp. 201–203 & *Journal of Peter Good*, p. 79.

drawings of animals. Brown apologetically confessed that he had "unpardonably neglected two opportunities ... from Madeira & the Cape of writing ... & now that a considerable & important part of the voyage is over, I am afraid I shall disappoint expectations, in not sending by the present conveyance, specimens & descriptions of what I have collected in Natural History." He was much too modest because his lengthy letter to Banks was filled with valuable information. Brown indicated that he had done "little in zoology" because of the time required to collect, preserve, and describe plant specimens. As far as mineralogy was concerned, he reported collecting rock samples that "presented itself on the surface." He explained that his efforts in transporting live plants were wanting due to the lack of proper facilities—i.e., a garden with boxes which could provide the plants "sufficient protection"—eventually resulting in the loss of approximately 70 species that were brought on board originally in good condition.[48]

Brown supplied Banks with some details of the events that occurred on the voyage up to that point and forwarded a small box of seeds to Banks with his letter. In his report, he included their encounter with Baudin's expedition and his visit with Baudin on the French ship of discovery, as previously cited, the well-equipped corvette, *Le Géographie*. *Le Naturaliste*, the other ship in the French party—also a corvette—was berthed in Sydney Harbour, and the only trained botanist on board was Jean Baptiste Leschenault de la Tour (1773–1826). The condition of these ships may have made an impression on Brown in comparison with their own leaky ship that they had become accustomed to. However, there was no mention of that in Brown's diary or in his letter to Banks.[49]

Brown also praised the work of Good, indicating that he was a "valuable assistant." He explained that Good did not have adequate quarters on board for taking care of living plants because the ship was not designed or suitable for the potting and boxing of living specimens—one of Good's duties in addition in his job as collector—so some of them had to be grown in the governor's garden. Brown was able to provide a summary of the numbers of plant species from King George III Sound they encountered in the 24 days they spent there: " nearly 500 species of plants." He indicated the amount of time they spent in each place, for example, "Kangaroo Island 5 days, King's Island 1 day etc."[50]

While they remained in the harbor at Port Jackson, the naturalists had the time to explore the region. Good went ashore with Brown, and Leschenault joined them. Brown was impressed by Leschenault, referring to him in his May 30th letter to

---

[48] Letter from Brown to Banks, May 30, 1802, original in *The Banks Letters*, British Library Add MS 32439, 61–64. Copies in Banks, *The Indian & Pacific Correspondence*, vol. 6, letter #56, pp. 69–71, and transcript in *Nature's Investigator*, pp. 204–206.

[49] Leschenault was a pupil of Antoine Laurent de Jussieu (1748–1836). Although De Jussieu's natural system of classification was not widely accepted in Britain, French naturalists readily embraced it.

[50] May 30th letter to Banks, British Library Add MS 32439 ff. 61–64, also Banks, *The Indian & Pacific Correspondence*, vol. 6, letter #56, p. 70.

Fig. 6.6 Government House in Parramatta, from the Mitchell Library, State Library of New South Wales

Banks as "an acute observer." Good reported that they "collected many fine plants."[51] Good gathered the seeds he collected from the southwestern coastal regions, and sent them to William Townsend Aiton (1766–1849)—Aiton was Superintendent of Kew, having succeeded his father who previously was the Royal Gardener at Kew Gardens—on the *Speedy*, a whaler bound for England. Brown continued to collect the plants from New South Wales, although initially the sparse entries in his diary give little notion of his daily ritual of work while the ship remained in Port Jackson. He took a walk from the Sydney west to Parramatta (on June 17th), what could be considered the beginnings of suburban sprawl, noting the plant life and landscape in his diary: "Walk from Sydney to Parramatta, cultivated ground about half of the way. Stumps of trees very numerous in all the farms seen—soil indifferent. Houses, offices [possibly outhouses or lavatories] &c. Tolerable situation of Parramatta. Pleasant." (Fig. 6.6)[52]

Brown used the time to catch up on his correspondence. The same day he wrote to Banks (May 30th), he wrote to Jonas Dryander and expressed disappointment that he had observed only 750 species of plants excluding cryptogams (but roughly 300 were undiscovered out of 750 observed). He also added that he needed decent writing paper because he lost part of his paper from the dampness of the place in which it is kept. He explained that what remained was "far from sufficient for the

---

[51] *Journal of Peter Good*, p. 79, entry for May 11, 1802.

[52] Diary entry for June 17, 1802, *Nature's Investigator*, p. 212.

remainder of the voyage" so he literally begged Dryander to "purchase" eight reams of "Imperial brown paper."[53]

Brown also wrote to Sir Charles Francis Greville (1746–1809), a close friend of Banks, a politician, an antiquarian, a collector, and a patron as well as a fellow of the Royal Society. Brown repeated what he told both Banks and Dryander: i.e., a summary of his observations, and about seeing both French ships, "Le Géographie at sea on the south coast, Le Naturaliste in Port Jackson, and having "picked up a little information concerning their past operations, which I have sent to Sir Joseph [Banks]." He indicated that the plants he found are mostly new species, somewhat contrary to what he told Banks and Dryander but speculated that "[James] Dickson will be sorry to hear that cryptogamic plants are neither numerous or singular; most of the lichens observ'd are well-known species; mosses are uncommonly few."[54]

Flinders wrote to Banks during their stopover in Port Jackson (May 20th), expressing his view about "the success of our voyage thus far." He reported some information he learned from the meeting with Baudin, particularly the topography and observations about the coastline and that "Port Phillip is surrounded by very fine country." He concluded by indicating "I say nothing of our scientific gentlemen, M[r] Brown being so much better qualified to tell his own storey than I am. It is fortunate for science that two men of such assiduity and abilities as M[r] Brown and M[r] Bauer have been selected, their application is beyond what I have been accustomed to see."[55]

Brown spent the months of May and June exploring the region around Port Jackson, sometimes with Good and Bauer, and sometimes alone. Good also went off on his own a number of times as well, for example, to Green Hills and Constitution Hill (June 2nd). Meanwhile, Brown went with Caley to North Rocks where he found nothing new (June 18th), but found Caley quite garrulous, noting that Caley used the generic names for plants but his Greek was "mostly bad." Brown thought Caley as a capable observer but his taxonomic knowledge was limited. Brown got along well with him and evidently was able to develop a working relationship with the difficult naturalist in spite of the latter's sometimes resentful attitude. The following day, he went with Caley again to Castle Hill, and found several species of *Eucalyptus* (including *Eucalyptus sclerophylla*, white gum), with most of the trees being *Casuarina suberosa* (a type of she-oak).[56]

On June 20th, Brown and his party (including Good) walked from Parramatta to the Hawkesbury River (referred by Brown as "the Hawksbury"). They took similar excursions in the next few weeks, along the various rivers in the area (the Nepean, the Grose, as well as the Hawkesbury) and various settlements. Brown noted that

[53] Brown letter to Jonas Dryander, May 30, 1802, *The Banks Letters*, British Library, Add MS 32439, 67–68. Also, a copy of the letter is in *Nature's Investigator*, pp. 206–207.

[54] May 30th, 1802, letter from Brown to Sir Charles Francis Greville, *Nature's Investigator*, pp. 207–209.

[55] Letter from Flinders to Banks, May 20, 1802. *The Indian & Pacific Correspondence*, vol. 6, letter #52, pp. 59–63, and there is a copy in *D.T.C.* 13: 114–118.

[56] Diary entries for June 18 and 19, 1802, *Nature's Investigator*, pp. 212–213.

the quality of the soil was quite good but very few unusual plants were observed although Good gave more information about the cultivated plants in the areas they explored. Because they were visiting farms and other areas that had been cleared for settlement, it was unlikely that they uncovered anything new or unusual there in the way of flora.[57]

The naturalists and their party returned to Sydney on June 25th, as preparations were under way for the ship's departure. Good was busy "securing former collections" and took "short walks in the neighbourhood," while Brown visited Botany Bay.[58] The *Investigator* was ready to sail and continue its circumnavigation of Australia. A makeshift greenhouse was constructed on the quarterdeck to keep the plants that Brown and others collected—eventually they were sent to Banks and the Royal Gardens at Kew. Flinders wrote: "Amongst our employments was that of fitting up a green house on the quarter deck … for reception of such plants as might be found by the naturalist, and thought worthy of being transported to his Majesty's botanic gardens at Kew." (They reduced the size for the safety of the ship.) Flinders continued, "Mr. Brown being of the opinion it would contain all the plants likely to be collected in anyone absence from Port Jackson …."[59]

Good reported prior to departure, he "took a short walk in the fields and collected a few things" in the morning.[60] Just before departing Port Jackson, Flinders noted, "Lieutenant John Murray, commander of the brig Lady Nelson, having received orders to put himself under my command, I gave him a small code of signals, and directed him, in case of separation, to repair to Hervey's Bay …."[61] On July 21st, 1802, the *Investigator* sailed northward from Sydney Cove, with the *Lady Nelson* serving as tender, i.e., as an auxiliary vessel for next stage of the journey, and it was frequently in sight of the *Investigator*.

---

[57] Diary entries for June 20–24, 1802, *Nature's Investigator*, pp. 213–215.

[58] *Journal of Peter Good*, p. 81.

[59] Flinders, *A Voyage to Terra Australis*. I, p. 231.

[60] *Journal of Peter Good*, p. 81.

[61] Flinders, *A Voyage to Terra Australis*, vol. II, p.1, entry for Thursday, July 22, 1802.

Chapter 7
The Crew Laboring Under the Same Disorder

Contents

Toward the Great Barrier Reef

The following day (July 22nd) the ship was clear of the coast and was again sailing in the open sea. Under a fresh breeze, they headed northward off the east coast of Australia, past Broken Bay (at the mouth of Hawkesbury River), Hunter River, and Port Stephens.[1] They remained at sea for a week until they reached Sandy Cape. There were few entries in Brown's diary but Good provided a detailed geological description of the area north of Sydney from observations of the coastal region he made from the ship: i.e., the coast was composed of granite with mostly calcareous soil.

On July 30th, they anchored near Sandy Cape (on Hervey Bay). Brown went ashore and found the fruits from *Pandanus odoratissimus*, an herb commonly known as screw pine, and the flowering spikes of a species of *Banksia*. Later in the day, he further noted "a small tree with a lamellate bark," *Melaleuca viridiflora*.[2] Brown went on to observe and collect 74 separate species of plants that are listed individually in his diary entry for July 31, 1802 ("Plants oberv'd Hervey's Bay, Sandy Cape").[3] Good noted some of the plants but did not compile a complete list.

---

[1] Flinders recorded, "In the morning of July 22nd, we sailed out of Port Jackson together {with the brig, *Lady Nelson*} and the breeze being fair and fresh, ran rapidly to the northward, keeping at a little distance from the coast," Flinders, *A Voyage to Terra Australis*. II, p. 2.

[2] The broad leaves of *Melaleuca viridiflora* are a good source of niaouli oil, a strong disinfectant used by Cook and which remains in use today.

[3] See *Nature's Investigator*, pp. 233–236, for the complete list and notes.

© The Author(s), under exclusive license to Springer Nature Switzerland AG 2021
J. Schwartz, *Robert Brown and Mungo Park*, Memoirs of The New York
Botanical Garden 122, https://doi.org/10.1007/978-3-030-74859-3_7

Good seemed to be more fascinated by the natives. Brown reported that they ran across a tomb containing the remains of several natives as well as some fishing nets that apparently were the equal of the European variety. Brown and other members of the party took the nets and left a hatchet and a red night cap in return. There is a slight discrepancy concerning the separate written accounts of Brown and Good because, according to Brown, after the nets were taken, there was further contact with the natives who were unarmed. The latter asked for their fishing nets back and the hatchet was taken back. In Brown's account, there was further negotiation and some nets were exchanged with the hatchets and red night caps. That seemed to put an end to further looting in Brown's account.[4] However, in his separate account, Good noted that Flinders personally put a stop to the removal of the nets.[5]

Brown noted the appearance and behavior of the natives: "In their features, colour &c. they are precisely similar to those of P[ort] Jackson. In size they appeared somewhat inferior. They were also less muscular but considerably plumper …. Many of them had long beards. All of them had the septum narium perforated. Some of them were painted with red ochre …. With respect to their language nothing was learned, except that it seemed to differ very considerably from that of the natives about Port Jackson."[6]

On August 1st, the ship weighed anchor in the Burnett River region and the following day they anchored briefly two miles from shore in Bustard Bay. The party moved on to Port Curtis, arriving there on August 5th. They explored the area extensively, and had a brief confrontation with natives who showered them with sticks and stones but retreated when Brown and others fired warning shots over their heads. Brown expressed some sympathy for them because he understood that he and his fellow shipmates were trespassing on the native soil of indigenous people, noting that they were basically unarmed: "Thus ended our first skirmish with these poor unarm'd savages, in which they seemd to have much the advantage of us in point of bravery & also in conduct."[7]

Brown did far better in terms of natural history, collecting a total of 49 angiosperm species from his and Good's excursions in the Port Curtis area, i.e., the southern end of Curtis Island, Facing Island, and the western shore of the mainland. Good explored Facing Island near Port Curtis where they first docked—discovered earlier by Flinders, in addition to Port Curtis which he initially named "Port I." They collected many plants, either in a dried condition or still alive which they planted in the greenhouse aboard the *Investigator*. His extensive plant list was inserted as a diary entry for August 5th.[8] He included in his list such plants as *Drosera indica*, a genus of insectivorous plant much admired by Darwin later on; a species of *Mimosa*; and

---

[4] *Nature's Investigator*, p. 231.

[5] *Journal of Peter Good*, p. 82.

[6] *Nature's Investigator*, pp. 231–232.

[7] *Nature's Investigator*, p. 238.

[8] In addition, they took samples of minerals for their collection preserved now in the Department of Mineralogy of the British Museum (Natural History).

broad-leaved tea trees.[9] Bauer drew a bandicoot, a rabbitlike marsupial, still found almost exclusively in Australia.

In the latter part of their visit, the ship traveled northward from Port Curtis in the narrows toward Curtis Island, bordered on its northern shore by Keppel Bay (August 9th–17th)—discovered earlier by Cook. They explored the northern end of Curtis Island (at Sea Hill), and found a few new species of plants such as *Suriana maritima*, a coastal cedar, several previously unidentified Proteaceae, and *Plumbago zeylanica*, an herb with medicinal properties commonly known as white leadwort. The bird life they encountered was unusual and varied, with ducks, teal, water hens, etc. During this phase of the expedition, Brown discovered the blue-winged Kookaburra (*Dacelo leachii*) but neither his nor Good's diary mentions this discovery.[10]

Often Brown and the others explored in separate parties and sometimes Brown stayed on board, studying and cataloging the plants that were collected. Frequently, Good hiked through the difficult terrain and collected new samples of flora. Occasionally, the roles were reversed, with Good remaining behind and Brown going into the field. Flinders led another party, sometimes with Westall acting as naturalist and draftsman. Bauer generally accompanied either Brown or Good but sometimes joined the Captain. On one such expedition (on August 14th when Brown was exploring Keppel Bay), Westall drew the view from South Hill, overlooking the southern shore of the bay.

Bauer drew mudskippers (jumping fish, *Periophthalmus modestus*), which fascinated Flinders. Banks previously noted these fish when he traveled with Cook. Brown observed that these creatures enjoyed an amphibian-like lifestyle, capable of staying on land for short periods of time: "We also observd here the small kind of fish mentioned by Capn Cook which leap among mud & stones & prefer leaping from stone to stone to taking the water."[11] Flinders also recorded his observations about this unusual fish: "The small fish which leaps on land upon two strong breast fins, and was first seen by Captain Cook on the shores of Thirsty Sound, was very common in the swamps round the South Hill." Brown did not directly speculate about a possible link between fish and amphibians but noted, "These creatures enjoyed an amphibian lifestyle."[12]

---

[9] The scientific names of many plants in this extensive list have been revised, *Nature's Investigator*, pp. 239–242.

[10] David J. Mabberley reported this based on the information he obtained from the "Bird Group" of the Natural History Museum at Tring, Hertfordshire (formerly the Walter Rothschild Zoological Museum), *Jupiter Botanicus*, p. 99.

[11] Quotation taken from Brown's diary in *Nature's Investigator*, p. 248.

[12] Flinders, *A Voyage to Terra Australis*, vol. II, p. 26, note for Saturday, August 14, 1802. Joseph Banks was the first naturalist to notice this unusual fish (on May 1779, in Thirsty Sound), *The Endeavour Journal of Joseph Banks, 1768–1771*. 2 vols. ed. John C. Beaglehole (Sydney: Trustees of the Public Library of New South Wales, 1962), vol. 2, p. 72. It is a description of an organism suggesting an evolutionary transition from fish to amphibians but Banks and the other naturalists at the time regarded this organism more of a curiosity than something that indicated mutability of species.

Before departing Keppel Bay, several other things occurred, one rather mundane that related to their investigations in natural history and the other that was somewhat worrisome at the time. On August 15th, while part of the crew had gone ashore having been granted liberty by Flinders near South Hill, Brown, Bauer, and Good took the opportunity to explore the South Hill area (described as "a little hill"), and discovered certain plants they previously had not seen including *Plumbago*. This excursion to South Hill proved to be quite worthwhile. To their delight, they also observed more examples of the mudskipper, a creature whose behavior they continued to enjoy.

However, at the same time, the rest of the crew had a brief skirmish with the natives. It was quickly resolved but in the evening it was discovered that several members—a petty officer and a seaman—had disappeared. Little could be done that evening but the next day (August 16th), along with friendlier natives who came "in considerable numbers to the beach" were "one of the Gentlemen & a Seaman who had lost themselves the night before." Brown added his description of the incident, making an observation about the appearance of the natives. "We had a friendly interview of them …. They were in general tall…& a few were uncommonly muscular."[13]

The next day (August 17th), they left Keppel Bay, passing Keppel Island, Cape Manifold, and finally the open sea on the way to Broad Sound and Port Clinton (originally named Port Bowen or Port II by Flinders), arriving on the 21st of August. Brown went ashore almost immediately after the *Investigator* docked, and he noted that the Norfolk Pine was rather reduced in size with a very ragged appearance and irregular growth. He also noted the geological composition—that the soil was composed of aggregate stone (fine granite and sandstone)—and there were a number of plants that Brown previously had not seen. There was also a good quantity of freshwater, a most favorable circumstance on such an expedition. On August 21st, Flinders described the scene: "A boat was dispatched with the scientific gentleman to the north side, where the hills rise abruptly and have a romantic appearance, another went to the same place to haul the seine [large fishing net] at a small beach in front of the gully between the hills, where there is a prospect of obtaining fresh water; and a third boat was sent to *Entrance Island* with the carpenters to cut pine logs for various purposes."[14]

Brown and Good went ashore the following day and continued exploration of the Port Clinton area. On August 23rd Brown remained on board initially, and then went to Entrance Island where he observed many pines. Flinders noted that the pines, *Araucaria cunninghamii*, resembled the pines of Norfolk Island. The ship's carpenter found this supply very useful: turpentine was oozing from between the bark and woody part of the trees and he busied himself in cutting wood for repair and replacement of some of the topmasts of the ship. Flinders noted that the pine

---

[13] Good identified the "Gentleman" as Thomas Evans who was a master's mate and eventually became a Lieutenant in the Royal Navy. *Journal of Peter Good*, p. 86 and *Nature's Investigator*, p. 248.

[14] Flinders, *A Voyage to Terra Australis*, II, p. 36.

logs were mainly employed to make a "sliding keel" for the *Lady Nelson*, which had been accompanying the *Investigator* as it sailed northward on orders from the Admiralty.[15]

With the ship stocked with sufficient provisions, they left Port Clinton the following day (August 24th), heading for the "Clara Group" of the Hervey Islands.[16] Although Brown found little variety of plants and nothing new, he did find the ripe fruit of a Rubiaceous plant with young berries that he had not seen before.[17] On the 26th the ships sailed through Strong Tide Passage—which really lived up to its name because a cutter was lost from these strong currents but fortunately there was no loss of any of the crew—past Shoalwater Bay. There they observed one of the highest points they had seen so far (2500–3000 ft), which Flinders named Mount Westall. They climbed Mount Westall to get a good view of the surroundings but found little vegetation. When they got back to the beach, the current of the tide was so fierce that there was some difficulty reaching the ship. Their cutter was cut adrift but the *Lady Nelson* saved the boat keeper. They spent the following day—(August 27th in Strong Tide Passage)—searching for the lost cutter but unfortunately their efforts were not successful.

On the 28th, they landed on Cape Townshend Island. Both Good and Brown did some exploring on the southeastern part, hiking up one of the hills and getting a good perspective of almost the entire island. They noted a number of mangrove swamps and numerous streams, but the soil was poor and sandy with very little plant diversity. They found blady grass (*Imperata cylindrica*)—a grass common in Australia's sandy regions—and a species of *Eucalyptus*, *Eucalyptus tereticornis*. Brown took various mineral samples, including quartz-sericite schist and white or milky quartz he called "pure fat quartz." In the meantime, Flinders explored the western part of the island, but again found no sign of the missing cutter. He noticed a few large bats called "flying foxes" and both he and Brown separately noted moderate-sized kangaroos.

The party spent the remainder of the month in the vicinity of Shoalwater Bay. On the 30th they did a considerable amount of walking in the Rocky Creek area where they encountered some natives who initially retreated from the strangers' approach. Later on, they tried communicating with the natives, who were well armed with spears of hard wood and wooden shields, but they were not hostile and seemed friendly. Both Good and Brown observed that they were moderate sized in stature, somewhat shorter than the natives they encountered in Keppel Bay. Because these natives had little contact with Westerners, Brown and his party found it difficult to communicate with them. Flinders, in the meantime, went on the *Lady Nelson* to

---

[15] Ibid.

[16] Brown and Flinders referred to the islands as "Harvey," not "Hervey" as Cook did.

[17] Rubiaceous plants belong to the Rubiaceae, a very large plant family often referred to as the madder, bedstraw, or coffee family. Examples of plants included in this family are gardenia, cinchona, and partridgeberry.

explore the eastern shore of Shoalwater Bay, i.e., the mangrove-covered flats intersected by tidal channels.[18]

Flinders found no freshwater and was unable to find any in the next few days spent in the area. Meanwhile, Brown remained on board working on the plant specimens that had been collected. On September 3rd, Brown, Good, and their party trekked up Peak Hill (Pine Mountain), and Brown described the scene as "the most beautiful we had seen in New Holland," quite an uncharacteristic statement from the normally reserved naturalist.[19] They found that the pines were not as large as those on Entrance Island in the vicinity of Port Clinton (Bowen or Port II originally), and observed a different species of *Eucalyptus*. However, for the most part, there were no unusual plants for the soil was relatively poor and sandy. Because their investigations consumed most of the day, they had to stay overnight in the area so they dined on fish and found sufficient freshwater to sustain themselves.

The following day (4th), they rejoined the ship and the *Investigator* left Shoalwater Bay, heading toward Thirsty Sound, and reached that destination the following day (September 5th). Brown reported that Good went on shore on the mainland and Bauer on the peninsula of Pier Head but "neither were successful in Botany."[20] On the 6th, Brown explored Pier Head and Flinders went to Tynemouth Island.[21] They made no new discoveries there. Because they were in rather shallow water (with shoals and a falling tide), they lost an anchor but some of the cables were recovered the following day.

Despite the shallow water, the *Investigator* sailed in Broad Sound and reached West Hill Island where they anchored at sunset on the 8th of September. They went ashore the following day and found the vegetation thick with large trees at the highest elevations, but little was accomplished, i.e., few things were observed that were new or noteworthy. On the 10th, Flinders pushed on to the Flat Isles further into Broad Sound. The Flat Isles had a greater variety of trees, including tulip plum (originally named *Spondias acida*, now called *Pleiogynium timoriense*) possessing large purplish fruit, and a medium-sized evergreen *Mimusops elengi* (called Spanish cherry), whose fruit was edible and had some medicinal properties. Then they sailed to the northern portion of Broad Sound and anchored "abreast" a "low round hill near the shore," i.e., Upper Head (September 12th). Brown reported finding a new "Myrtus with a flower bud" (actually *Austromyrtus racemulosa*, the Python tree).[22]

They remained in the Broad Sound area for almost a week but the shallow waters hindered the Captain's explorations. On the 13th the *Investigator* became grounded in shallow water so Flinders decided to continue surveying the area on the *Lady Nelson*—located a half a mile away—in addition to taking readings on a whaleboat.

---

[18] *Nature's Investigator*, p. 257.

[19] *Nature's Investigator*, p. 261.

[20] *Nature's Investigator*, p. 263.

[21] Vallance et al. described Pier Head as an andesitic plug of the Cretaceous geologic age, i.e., volcanic igneous rock composed of silica, *Nature's Investigator*, p. 264.

[22] *Nature's Investigator*, p. 269.

The following day (14th), the *Lady Nelson* also became grounded. Unfortunately, in taking the ground, the main keel was broken. When the tide came in several hours later, the ship was freed but the main keel needed to be repaired. This new problem would have to be addressed but meanwhile the *Lady Nelson* would not be as serviceable as before as an exploring vessel.

On the 15th, Brown and Flinders walked to "The Brothers," a group of low hills near the head of Herbert Creek, a small channel in Broad Sound. Brown observed "several plants I had before found, among others *Marsilia quadrifolia* [probably *Marsilea hirsuta*, a short-fruited Nardoo] in great abundance but without fructification." Nardoo, commonly called rough water clover, resembles water lily but actually is an unusual type of fern, common to the wet areas of Australia. Early in the nineteenth century, the term fructification was not just reserved for flowering plants (angiosperms) but was also applied to spore formation in ferns and mosses.[23]

The next few days, there were continued attempts at exploration but they were limited by the constant grounding suffered by the *Lady Nelson*. Eventually, they boarded the whaleboat and returned to the *Investigator*, and Brown spent the last day in the area (at Upper Head on the 18th) describing and studying the plants he found at the head of the Sound and Good had found at Upper Head an area where they had pitched tents. While the *Lady Nelson* was being repaired, Brown sailed with Flinders on the *Investigator* to Thirsty Sound and anchored off of the Mangrove Islands. Brown spent the next few days exploring the mainland and Mangrove Islands off of Thirsty Sound, taking little in the way of specimens, but did note a number of species of mangrove trees (*Rhizophora*) as well as the "blood red cliffs." Flinders found a passage that was connected to Thirsty Sound in a whaleboat around a small island, called "Long Island," thus adding to the information that Cook gathered there years before.

On the 24th, they sailed across Broad Sound to Upper Head, allowing Brown to conduct further botanical exploration of the area. He reported adding a few plants to his collection and a bronzewing pigeon (probably the common bronzewing), birds native to Australia (26th). On the 27th, they departed Upper Head and sailed to the Northumberland Islands in the Percy Isles group. They anchored at a small island, Pine Island, and Brown noted the variety of the shrubs and trees as he hiked with Westall, who sketched while Good remained behind on the ship. On the 30th, they went ashore on Middle Island, and Westall sketched the palms they encountered and *Flagellaria indica*, a monocotyledonous herb related to corn and sometimes eaten by the natives, commonly referred to as whip vine, bush cane, or supplejack. On October 2nd, they landed on the largest of the Percy Isles, now called South Island. Brown explored the island with Flinders, noting *Calophyllum inophyllum*, a large evergreen tree sometimes called ballnut while Good remained behind except to gather some soil for the plants they had brought on board. There was some freshwater but in limited supply, and little in the way of flora that was unusual.

---

[23] *Nature's Investigator*, pp. 271–273.

Brown remained on board the following day, while Bauer and Flinders journeyed to Pine Peak Island. They brought back *Morinda citrifolia* (Indian or beach mulberry, a small tree with flowers and fruits the entire year), and *Scaevola taccada*, commonly known as half-flower shrub, found in Hawaii and many Pacific Islands as well as in Australia. Natives sometimes used it as a medicine (but it seemed toxic to fish). After taking on enough freshwater for the next part of their journey, they got everything on board and headed for the Great Barrier Reef, reaching it the following day (October 4th).

Brown was fortunate in that Captain Flinders took an active interest in the scientific work Brown was engaged in. He assisted him and the other naturalists whenever he could. It was helpful that Flinders shared Brown's interests. He took part in collecting and added his own perspective to the observations and the work being conducted in natural history. This proved to be quite an asset for Brown.

The Great Barrier Reef

Brown's initial glimpse of the vast coral expanse of the Great Barrier Reef did not lead him to make great exclamations of wonder or show unusual excitement in the same manner later naturalists did when they first encountered tropical or other settings they had not observed before. In South America, Darwin was taken by the lush tropical setting with its abundant variety of life, and in Madeira, in parts of Australia, and in New Guinea, Huxley was equally moved by the beauty of the natural world he observed.[24] Captain Cook discovered the Reef in 1770. He was the first Westerner to do so because Magellan had missed Australia. Cook made dramatic contact with the vast reef when the *Endeavour* hit it with such force that the ship remained aground for some time and took a great effort to extricate it; one of its cannons remained imbedded in the coral.

Brown reported the first sighting of the "reefs" in his October 4th diary entry but because navigating was exceedingly difficult there, they were not able to explore at first. Finally, on the 9th, Brown accompanied Flinders in the whaleboat to one of the reefs, and he noted the presence of Holothuria (sea cucumbers), sea urchins, as well as common varieties of coral. He reported the "gigantic cockles of Capn Cook in considerable abundance," the giant clam, *Tridacna gigas*, which were eaten by the crew, and were universally disliked because they produced nausea (far different from what happened on the *Endeavour* years earlier).[25] Flinders reported the

---

[24] Despite popular belief, Darwin never spent time at the Great Barrier Reef—his visit to the Cocos Islands (Keeling Islands) was the inspiration for his treatise on coral reefs—and Huxley, when aboard the *Rattlesnake*, spent much more time in the Torres Strait area, separating the Cape North Peninsula of Australia from New Guinea so he conducted little dredging of specimens from the Great Barrier Reef.

[25] *Nature's Investigator*, pp. 283–284.

problems they had in negotiating the narrow passages in the reefs; the *Investigator* lost an anchor and the *Lady Nelson* also lost one and broke another.[26]

With some skillful maneuvering, they were able to reach the Cumberland Islands on the 15th and anchored on Calder Island. The following day (16th), Brown went ashore and he reported seeing some plants not seen before but were "imperfectly known," including *Cycas circinalis*, a cycad not native to Australia but which Flinders observed in Keppel Bay and that Brown was familiar with from his study of Banks's Indian herbarium. Brown reported seeing it with its "fruit and young male flowers," but these were actually the female and male cones. Coniferous plants were still considered flowering plants and the distinction between the two different groups of seed plants was not established until Brown and others made major strides in accurately classifying these plants.

Brown also observed *Dombeya norfolkiensis* (actually *Araucaria cunninghamii*, hoop pine) "in considerable quantity," *Casuarina equisetifolia* (she-oak or Australian pine), and *Morinda citrifolia* (Indian or beach mulberry). He collected *Crinum venosum*, a lily-like perennial.[27] While Brown was concentrating on collecting and describing the varied plant life encountered, Flinders was much moved by the site of the coral reef. His report of the sight of the vast expanse of coral was effusive, in contrast to Brown's depiction; Flinders described the coral reef as presenting "a beautiful piece of marine scenery."[28]

Because the *Lady Nelson* was going to return to Port Jackson, Brown decided to use the opportunity of getting letters back home by writing to Banks to inform him of his progress so far. In his letter of October 17, 1802—with the heading "H.M.S. Investigator, at sea, coast of N.S. Wales, near the Cumberland Isles"— Brown provided a list of where they had anchored and his success in examining "numbers of plants … either with flower of fruit does not exceed 450, including grasses but exclusive of cryptogamic plants …. A very small proportion are new, most of them having been discovered by you." He went on to write that his zoological "acquisitions" were "by no means great" but "the mineralogy of the east coast" was more varied than the south. He also indicated that he was growing short of paper and fearful that he would not be able to replenish the supply when he got back to Port Jackson so he again requested more paper, a request he made earlier. He closed by sending his best to Greville, Dryander, Correa, and Dickson. It was a short letter, considering all that he could have reported because the *Lady Nelson* departed the next day (October 18th) and he had to get it off as expeditiously as possible. Brown wrote in his diary that Flinders was sending the ship back because of "dull sailing, her great leewardness [that is, moving in a direction away from the wind] & more especially her being reduced to a single anchor."[29]

---

[26] Flinders, *A Voyage to Terra Australis*, II, pp. 99–101.

[27] *Nature's Investigator*, pp. 285–286.

[28] Flinders, *A Voyage to Terra Australis*, II, p. 94.

[29] Brown letter to Banks of October 17, 1802, is reprinted in its entirety, with footnotes, in *Nature's Investigator*, pp. 287–288, and a copy is in the *D.T.C.* 13. ff. 277–278.

Torres Strait, Queensland

They spent the next day (19th) trying to negotiate the difficult waters in the Great Barrier Reef, finally finding passage on the 20th to the open sea. After a week on the open sea, the *Investigator* was in sight of the Torres Strait—a body of water between Cape North, Australia, and the Melanesian island of New Guinea—and they arrived at the Murray Islands on the 29th. The Strait consisted of reefs and islands, making it difficult to navigate. Brown noted that the islands were densely populated, and soon natives paddled to the ship in long and narrow canoes. Captain Flinders and others on the *Investigator* exchanged goods with the natives, accepting bananas, cocoa nuts, bows and arrows, various ornaments, etc. for iron. It appeared that the natives—who were friendly and not fearful of the Westerners—were primarily interested in iron; their word for it was "Toure."

Brown made an extensive list of the plants he observed from the Murray Islands, Zuizin Island, and Halfway Island, such as *Guettarda speciosa*, beach gardenia, *Suriana maritima*, bay cedar, *Capparis lucida*, coast caper, *Sophora tomentosa*, necklace pod, various species of kelp (*Fucus*), lichens, and many other plants indigenous to the area.[30] Although Brown encountered few new species, his observations and collecting were quite productive as he developed several extensive lists of plants and occasional animals.

On the 2nd they weighed anchor at Goods Island—in the Prince of Wales group in the Torres Strait—and Brown conducted a thorough investigation of the area, resulting in his extensive plant list (contained in his diary entry of November 2, 1802). It included a number of the species he listed on October 30th from the Murray Islands such as *Guettarda speciosa*, *Capparis lucida*, and different ones such as *Hibiscus tiliaceus*, a large shrub or tree native to eastern and northern Australia, and *Thespesia populnea* (the Portia tree). Brown also noted a large red cedar (*Toona ciliata*), and that the hills were "barren and stoney" and composed primarily of granite with "small veins of quartz" ("small crystals of quartz").[31]

The following day (3rd), the *Investigator* sailed into the Gulf of Carpentaria, a large shallow sea bounded on three sides by northern Australia, and on the north by the Arafura Sea, which lies between northern Australia and New Guinea. The first Westerners who explored this area were the Dutch in the seventeenth century. The water was quite shallow, so they did not stop for any appreciable time until the 7th, when they anchored three miles from shore in the Pennefather River region of Cape North (originally called the Coen River, in the western part of Cape North). Brown accompanied Flinders on the whaleboat and they landed at the northern shore of the estuary. They observed natives through spyglasses, noting that they were armed with woomera, an armed stick used to throw spears, but there was no incident reported. Brown collected many plants and again compiled an extensive list including *Capparis umbellata*, caper bush, a plant whose fruit has some medicinal

---

[30] See *Nature's Investigator*, pp. 293–295 for the complete list.

[31] *Nature's Investigator*, pp. 297–300.

properties; *Sesuvium portulacastrum*, sea purslane, a perennial, succulent herb; *Exocarpos latifolius*, native cherry, an evergreen shrub; coastal she-oaks (*Casuarina equisetifolia*), etc.[32]

The next week or so was spent sailing along the coast of the Gulf of Carpentaria region, heading south and west and then north, along this U-shaped body of water. On November 16th they arrived at Sweers Island of the South Wellesley Islands. They went ashore the next day but initially Brown observed nothing unusual, save for some coral, and birds like *Charadrius* (a genus of dotterels or lapwings from the plover family), black cockatoos, and white cockatoos. The discovery of rotten timbers on the ship may have been the most noteworthy event, accounting for why the ship's leaking had increased. Flinders decided that he would have to remain at Sweers Island to see if it would be possible to repair the defective wood. Good went ashore on nearby Bentinck Island and Brown went first to explore Sweers (on the 18th), and then on to Allen Island (on the 20th), but accomplished little in the way of natural history. During the following week, from November 22nd to 28th, many more defective timbers were discovered on the *Investigator*. It was obvious to Flinders by this time that the *Investigator* was going to have to undergo more extensive repairs.

At the beginning of December, they continued the exploration of the Gulf, with Brown noting that Mornington and the Sydney Islands differed from the "published charts" (December 1st–2nd). The following day, they were at Bountiful Island and the Wellesley Islands—sometimes referred to as Turtle Island for the large amount of turtles they observed in the water. The island was relatively bare, with only 65 plant species, although Brown noted *Pisonia grandis* (grand devil's-claws, a tree found in the tropical regions of the Pacific), a plant he "had not seen before." They also observed a good number of sharks.[33]

On December sixth, they anchored off Pisonia Island and Brown went ashore. Pisonia was "a small rather low & exceedingly well wooden island," and its dark green foliage led them to deduce that the soil of the island was quite fertile. Flinders and Brown went from the ship to the island in the whaleboat. Brown reported finding *Guilandina bonduc*—the genus is named *Caesalpinia* today—a vine commonly known as gray nicker. He also observed and recorded a number of birds including the Australian bustard, *Cuculus spectabilis* (a cuckoo), *Otis oedicnemus* (the beach stone-curlew), a species of heron, a species of pigeon or dove, and possibly a falcon.[34] After spending another day off of Mornington Island, they sailed to the border of Queensland and Northern Territory. On December 10th–11th, they anchored near the Calvert River, where they were grounded for a short time, and then briefly spent a day near the mouth of the Robinson River (12th).

---

[32] See *Nature's Investigator*, pp. 302–307, for a lengthy account of the exploration in the Gulf of Carpentaria region.

[33] *Nature's Investigator*, p. 319.

[34] *Nature's Investigator*, pp. 320–321.

Northern Territory

The *Investigator* and its crew left the area, moving on to the Sir Edward Pellew Group in the southwest corner of the Gulf of Carpentaria (named after Admiral Edward Pellew, 1757–1833, 1st Viscount Exmouth, who fought in the American Revolution and the Napoleonic Wars). They landed on Vanderlin Island on the 15th but spent less than an hour there before they moved on to Cabbage Tree Cove on North Island. Brown explored the island with Flinders, and Good went ashore the following day (17th) while Brown stayed on board. Good collected "several fine plants," including "an elegant *Sterculia*, a tropical chestnut, and a fine blue water Lilly," *Nymphaea stellata*.[35] On the 18th, both Brown and Good first returned to North Island, and then went on to Vanderlin Island the following day where Brown noted a nutmeg, *Myristica insipida*. On the 20th, the *Investigator* remained off of Centre Island while Brown returned to North Island and explored this ground on the 22nd (Cabbage Tree Cove). Despite not finding anything of a startling nature, Brown and Good explored another part of North Island (near Macassar Bay on the 23rd), where they observed an acre or so of mangrove (*Rhizophora angulata*) and a species of *Pandanus* (screw pine). They also observed traces of the natives, discovering that the aborigines were active in boatbuilding. They continued exploration of North Island and Vanderlin Island through Christmas Day, and then Good went with Flinders to Observation Island where he gathered some plants for the "Garden." Although they did not gather a great deal of new information in terms of natural history, "these fine islands" as Good described them provided "excellent shelter, wood & water."[36] What they had not accomplished in natural history, surely their efforts were useful in describing the region's geography.

They spent the remainder of the year at sea, before arriving at Maria Island ("Cape Maria") shortly before New Year's Day. Brown and Good went ashore on January 1st, 1803, and found two or three plants that they had not observed before but his diary does not indicate what they were and his plant list for the island has not been located. Several days later (3rd), a wind filled the whaleboat with water and two sailors went to steady it. Unfortunately, conditions became more severe, and the result was that one man, William Murray, died by drowning in the accident.[37]

The *Investigator* anchored off the mainland in the Gulf and southwest of Bickerton Island on the 4th. Brown and Good went ashore and surveyed the area, finding a freshwater lake, a great number of plants—nearly 200, 26 of which Brown had not seen before—a number of animals, including small kangaroos, and birds like a variety of ducks and black crows, as well as human skeletons in hollowed-out trees. Good appeared from his account to be more fascinated by the unusual method that the aborigines buried their dead than the abundance of plants they found but Brown made the most of this stopover by collecting samples to bring with him

---

[35] *Journal of Peter Good*, p. 106.
[36] *Journal of Peter Good*, p. 108.
[37] A detailed account is in Flinders, *A Voyage to Terra Australis*, II, pp. 181–182.

although no list of them has survived. Flinders spent the next week or so (January 5th–13th) anchored off of Groote Eylandt, conducting an extensive survey of the place. Brown did not go ashore nor did he make any entries for this period of time.

On January 14th, Brown went ashore on Chasm Island but found very few plants and hardly anything that he had not seen before on this very rocky island with its abundance of *Myristica dactyloides*, a nutmeg used as a medicinal herb (that is now on the endangered species list), and what he identified as *Eugenia pomifera*, which was probably *Syzygium suborbiculare*, belonging to the myrtle family commonly referred to as Lady Apple whose fruit is the size of an apple. Flinders observed paintings the natives drew in predominantly yellow ochre color on the rocks, and Westall made copies of them. Brown explored Groote Eylandt on the 15th, observing *Acacia*, thorn trees or wattles, and brought back some samples of the rock from the island including fine-grained reddish sandstone and a coarser grained rock containing quartz pebbles. The following day, Brown remained on board while Good went with Bauer and Allen to survey Winchelsea Island where they found little they had not seen before. On the 18th the *Investigator* landed off of Bustard Island, aptly named for the number of Kori or Australian Bustards they observed there—a large terrestrial bird, similar to the turkey.

Brown explored a very stony island the next day with Bauer—initially labeled "island r" on the charts, and then named Burney Island after Captain James Burney (1750–1821)—and they collected some samples of rocks (sandstone), observing *Eucalyptus* and *Melaleuca*, a myrtle sometimes called paperbark. The *Investigator* then proceeded to Morgan's Island in Blue Mud Bay and Brown went ashore with Good (21st) but little natural history was accomplished there because of a tense encounter with the natives. John Whitewood, the master's mate, tried to establish contact with the aborigines but they proved to be quite aggressive, resulting in Whitewood receiving six separate wounds from the spears, injuries he fortunately was able to recover from. There were further problems as Benjamin Morgan, a marine, was overexposed to the sun while gathering wood on the island and unfortunately passed away in the evening after returning to the ship. They left Blue Mud Bay and traveled to Cape Shield, north of Blue Mud Bay arriving there on the 26th.

At Cape Shield Peninsula, Brown and Good went ashore and found a freshwater lake and observed *Vallisneria nana* (freshwater eelgrass), several species of charophytes (green algae, possibly muskgrass), and *Najas* (water nymph). They saw a great number of teal and a few ducks as well so their food supply was replenished. They moved on to Grindall Bay, a northern extension of Blue Mud Bay. Brown went ashore with an unidentified sailor, and they lost sight of the ship. Worried about Brown and the sailor, Flinders had a shot fired off to give Brown an idea where the ship was. As it was, Brown and his companion had to wait until the morning to rejoin the ship. After Whitewood's experience, Flinders became quite concerned, and Brown was worried as well as his diary entry indicates.[38] Fortunately, both

---

[38] *Nature's Investigator*, pp. 350–351. Brown realized that they would have to spend the night there. He was "much fatigued" and was burdened by "boxes & paper books." The sound of the

Brown and his fellow crew members were able to rejoin the party the next morning. At the same time, Good explored—with Allen—the area of Point Blane but did not find much in the way of new plants. He noted a large marsh of freshwater, "much frequented by" ducks and cranes and several emus.[39]

The *Investigator* sailed for a few days in the vicinity of Point Blane and finally left Blue Mud Bay on the 1st of February, arriving at Caledon Bay on the 2nd. They anchored in Caledon Bay on the 3rd and they prepared for a week of thorough investigation. The natives were friendlier and Brown was fascinated with the "fish gigs [spears for fishing] & womeras."[40] On the 4th, Brown, Good, and Westall began their exploration of the area, making certain that they were in sight of the ship. Because the natives were quite friendly, they were invited to accompany the naturalists and their party. A mangrove swamp impeded them so they had to change direction. In the meantime, some of the natives made off with a hatchet, and then a musket. It took them some time to get that sorted out so their first day was largely wasted from the standpoint of meaningful work in natural history.

The next day, they explored the area near Mount Caledon and became friends with one of the natives, Margandee (often referred to as Margandine). Brown's diary for February 5th contains an extensive list of the names of the natives, the plants they observed with the native words for them, and the native words for parts of the body. The aborigines fascinated Brown (as well as the others), so much so that they were spurred into making ethnological observations in addition to their main object of study of the flora and fauna of the region. Some of the plants noted by Brown in the Mount Caledon region included *Gardenia megasperma* (bush gardenia), *Boerhaavia diffusa* (red spiderling), *Bruguiera gymnorhiza* (black mangrove), *Avicennia marina* (gray mangrove), and *Sesuvium portulacastrum* (sea purslane), and several different types of grasses and a number of birds such as the collared kingfisher.

On the 6th, Brown accompanied most of the party and climbed Mount Caledon where they were quite taken by the marvelous view while Good gathered a box of plants "for the garden."[41] Brown did not refer much to the plants discussed by Good in his journal but he mentioned the discovery of freshwater. A beautifully colored parrot (or parakeet), *Platycercus venustus*, was discovered and shot—it was drawn by Bauer later on—and is now known as Brown's parakeet or rosella or Northern rosella.

An ominous development was that some of the crew began to grow sick with diarrhea and fever; it proved to be a most unfortunate development that later would pose a serious threat to the work Brown was conducting and to the entire expedition itself. On the 8th, Brown went in a yawl with Good, Westall, and Allen to "botanize"

---

guns from the ship indicated where they had to walk the next morning, and how far away the ship was, p. 350.

[39] *Journal of Peter Good*, p. 113.

[40] Womeras were weapons for spear throwing, *Nature's Investigator*, p. 352.

[41] *Journal of Peter Good*, p. 115.

the head of Caledon Bay. Their plans were somewhat thwarted by a skirmish they had with the aborigines. The next day, Brown, Good, and Allen explored the opposite side of the Bay and collected over 40 plant specimens, including *Leea rubra* (red leea) and *Myristica insipida*, Queensland nutmeg.

The North Coast

On the 10th, they left Caledon Bay but not before Brown visited Dudley Island, a small isle where he found nothing new or unusual. They were at sea for several days and then went to Gove Harbour in Melville Bay on the North Coast. They anchored in North Arnhem Bay on the 13th, and Brown went ashore where he observed some grasses such as *Thuarea involuta*, bird's beak grass, and *Daphne indica*, winter daphne, another medicinal plant although its effect is not indicated. They spent several days at Half Tide Point and Drimmie Hill where Brown observed *Bombax ceiba*, cotton tree, in large numbers, and a number of birds such as pheasants, white pigeons, and Northern Rosella (or Brown's parakeet). On the 16th Brown landed on Strath Island in Melville Bay, and he accompanied Flinders in a small boat to explore the distant parts of the Bay. He saw *Sonneratia alba*, the mangrove apple, for the first time.

The following day (17th), they met Malay fishermen in the English Company Islands who were in six small vessels. Brown expressed some surprise when he spotted them because initially he observed that they were flying the Dutch flag. Flinders sent his brother, Samuel Flinders, to ascertain what they were doing on the northern coast of Australia, and he learned that they were part of a fleet of 60 proas (or prows, Malay vessels) who were sent by their leader, Raja Bone, to fish for trepang, edible holothurians (sea cucumbers) that they sold to the Chinese. Captain Flinders sent a request to the commander of the proas to come aboard the *Investigator*. The Malays were also interested in what the *Investigator* was doing in these waters and were surprised to learn that there was an English colony in New Holland. Although they enjoyed the wine that was offered, they declined to dine on the *Investigator*. Nevertheless, they parted on friendly terms and several gifts were exchanged.

Flinders was able to negotiate a difficult passage between a previously undescribed cape and several offshore islands. Eventually, he named the cape, Cape Wilberforce, after his good friend and supporter, William Wilberforce (1759–1833), a man best known for his strong abolitionist views (of slavery).[42] The islands were

---

[42] William Wilberforce (1759–1833) was a politician who became a follower of the evangelical branch of the Church of England. His religious commitment drove his interest in creating a more moral society. As a member of parliament, he worked to end the slave trade and abolish slavery. In 1807, he led efforts in getting parliament to pass the Slave Trade Act which ended the slave trade. Poor health made him resign from parliament but he continued his work to abolish slavery com-

named the Bromby Isles after the Reverend John Healey Bromby (1771–1868), a resident of Hull, England.

The following day (18th), Brown accompanied Flinders back to the "proa" to get more information concerning the Malay fishermen, i.e., "for the purpose of more fully satisfying ourselves concerning the object of their voyage."[43] On the 20th, Brown landed on Cotton Island where he and Good explored a thickly wooded valley and observed *Myristica insipida* (Queensland nutmeg), *Ficus racemosa* (cluster tree), *Terminalia muelleri* (Australian almond), and *Hibiscus tiliaceus* (sea hibiscus or coastal cottonwood). The following day, they landed on Pobassoo Island where he found little that was of any value, and on the 22nd, Good went ashore on Astell Island where he also did not find anything new or of particular interest. This pattern repeated itself the next few days; Good went to Pobassoo Island on the 23rd and found nothing unique; Brown went ashore on Inglis Island, the largest island in the group; and Good visited Bosanquet Island the following day (25th). Finally, Brown and Good revisited Inglis Island on the 26th but again there was nothing new that either man reported. The *Investigator* weighed anchor at Malay Road, and it sailed near Mallinson Island on the way toward Arnhem Bay.

On March 1st Brown went ashore on Mallinson Island where he found two new plants, "a commelina [and] a syngenaesious plant [where the stamens were fused together by the anthers] but not easily referable to any known genus …."[44] In Arnhem Bay on the 2nd Good went ashore but Brown had to stay behind because of a sore on his foot.[45] Good returned the following day, finding *Osbeckia chinensis* (an herb still used to treat diarrhea), and a new species of *Embothrium* (likely *Grevillea*), belonging to the Proteaceae.

On the 4th, Good and Allen went ashore at Everett Island where the ship docked but they found few new plant specimens but they did find one remarkable fish that resembled salmon. Because salmon was not found in Australian waters, the fish was probably threadfin, belonging to the Polynemidae, likely *Eleutheronema tetradactylus* (also called Cooktown or threadfin or blue salmon). The survey of the coast of Australia was about over although there were a few additional excursions in the vicinity of Arnhem Bay. They remained at sea for almost the rest of the month, heading toward Timor and eventually to return to Sydney.

---

pletely. Ultimately, his goal of completely abolishing slavery was successful when parliament passed the Slavery Abolition Act in 1833 3 days before he died.

[43] *Nature's Investigator*, p. 373.

[44] *Commelina* is a genus of approximately 200 species commonly referred to as "dayflowers." *Nature's Investigator*, p. 379.

[45] T.G. Vallance et al. suggested it likely that Brown's sore foot was due to "scorbutic ulcers" from scurvy, *Nature's Investigator*, p. 380.

Timor and Then Sydney

Because of the growing cases of scurvy, Flinders decided to go to Kupang in Timor (Malaysia) to take on fresh supplies. He may have been suffering with some symptoms of scurvy and there was concern about how long the temporary repairs that were done on the ship would last. The decision to go to Timor eventually proved to be disastrous because in addition to replenishing supplies, they evidently brought an infection on board, a form of tropical dysentery, which eventually claimed many lives including Good's.[46]

On March [30] 1803, while the *Investigator* was in sight of Timor, Brown composed a long letter to Banks, describing the events that had occurred since he last wrote while they were docked in Sydney. He explained that "the crazy state of the ship" was the reason that they were abandoning their survey of that region of the northern New Holland coast (e.g., shortage of provisions, the state of the *Investigator*, the health of crew and officers). He indicated that in spite of the setbacks, they had been able to explore the Gulf of Carpentaria and "Arnhem's land." Now they planned to make their way back to Port Jackson. The tone of this letter was quite pessimistic, expressing disappointment with the work he had done so far.

Brown was unnecessarily negative. He had accomplished a great deal although his results in the northern territories pale in comparison with the success he had in Victoria. He felt that his botanical collections were fewer than he had hope for: "the number of species in addition to those we had already observed in other parts of the coast & excluding Grasses & Cryptogamic plants scarce amounting to 400." He reported that they had not done much in zoology for there was not much to do, noting he "met only with" quadrupeds like kangaroos, and Australian dogs, and their "additions to Ornithology are exceedingly few." He also described their encounter with Malay fishermen (from Celebes), and that a fleet of about 60 vessels ("proas") set off annually to collect a marine animal, trepang (a holothurian). Mineralogy continues "as barren a field as ever." He ended his letter by asking to be "remembered" to Dryander and Correa.[47]

The next day (31st), they arrived in Kupang, Timor, and noted a Dutch brig and an American ship that had just arrived from Europe. Good noted that the island was "peculiarly fertile in comparison to anything we had seen during this voyage." When Flinders and Brown went ashore to meet the governor (on April 1st), Brown was given permission to "botanize" the island.[48] The governor warned them not to venture any further than 20 miles in any direction into the country because there

---

[46] On Cook's first voyage, Banks's artist, Sydney Parkinson, died in Timor from dysentery (1771). David Nelson, Bank's collector later on (1789), also succumbed from the same illness, ironically after he had survived the mutiny on H.M.S. *Bounty*.

[47] Letter from Brown to Banks, March [30] 1803, British Library, Add. MS 32439, ff. 89r–90v and *D.T.C.* 14: 22–26. Also in Banks, *The Indian and Pacific Correspondence*, vol. 6, Letter # 104, pp. 153–155.

[48] *Journal of Peter Good*, p. 121 and *Nature's Investigator*, p. 389.

could be problems with the natives. Brown and Good did not take any chances, confining their surveys to the neighborhood of Kupang. Along with his diary entry for April 1, 1803, Brown included a list of 318 plants that they observed, supporting Good's optimism about the fertility or abundance of life on the island. The plant list included five species of *Scirpus*, belonging to the Cyperaceae (sedges); eight species of *Panicum*, from the Poaceae (grasses); *Pontederia cordata* (pickerelweed); *Artocarpus heterophyllus* (jackfruit, a tree from the mulberry family); and *Phaseolus lunatus* (the cultivated leguminous Lima or butter bean). Brown's diary entry for April 6–7th indicates that they were successful, observing nearly 300 species of plants although their excursions "were limited to the neighbourhood of the town."[49]

Before the *Investigator* left Timor, Brown wrote to Charles Francis Greville (sent out on April 7, 1803), and described what had occurred since he last wrote to him when they were in Sydney. His letter (dated March 30th) essentially repeated what he had written to Banks, and added his wish to be remembered to his "good friend Dickson."[50] They left Timor on the 8th and spent several weeks at sea. Brown's entry for April 20th mentioned that there were symptoms of dysentery (for the first time), and Good also noted "many of the crew complaining of disorders & dysenterie."[51]

The ship was mainly at sea while they searched for Tryal Rocks, the scene of a shipwreck in 1622 of the *Tryal* of the English East India Company. They were unsuccessful in these efforts. Flinders was unable to find Tryal Rocks, as other expeditions had before him in attempting to find this site or landmark. The sickness was growing worse on board so Flinders was determined to get to Sydney as soon as possible. Meanwhile, Brown busied himself with the plants he collected. Good's diary noted in his last entry (May 17th) "past noon departed this life Charles Douglas Boatswain of Dysentery."[52]

Good survived the voyage but succumbed to dysentery and died on June 12th while the *Investigator* was docked in Sydney Cove, a tragic end for a very talented investigator who made a major contribution to botany with his indefatigable work in collecting and observation. In addition to the seeds he collected and his many observations, his diary serves as an invaluable record of the expedition. The last leg of the trip—in the northern reaches of New Holland—did not turn out as well as the previous part of the trip in southern Australia (before they visited Sydney for the first time), and they were hard hit by the outbreak of disease. However, Brown's resistance to the disease that hit many members of the crew apparently was better than people like Good and this enabled him to withstand the rigors of exploring in uncharted territory.

The *Investigator* spent the rest of May and the first few days of June at sea with a few brief stops at Cape Leeuwin in Western Australia, King George Sound, Middle

---

[49] *Nature's Investigator*, p. 402.

[50] *Nature's Investigator*, pp. 388–389.

[51] *Journal of Peter Good*, p. 121 and *Nature's Investigator*, p. 403.

[52] *Journal of Peter Good*, p. 122.

Island in the Archipelago of the Recherche, finally reaching Sydney on June 9th (1803). The crew and the captain and other officers were quite weary but overjoyed that they had arrived and anticipated that they would be in port for some time given the condition of the ship and the health of the men on board. No one then realized how long the layover would be and how the fate of the badly damaged *Investigator* would drastically affect the rest of the voyage as well as play a significant role in determining the fate of Captain Flinders.

Chapter 8
Mungo Park's Last Journey

Contents

Park's Life at Home

While Brown was exploring Australia with Flinders, Park was not idle. After he returned from Africa in 1798, he stayed on in London, working on his notes and other material in preparation of writing an account of his eventful journey in Africa. He arrived in London rather unceremoniously on Christmas Day, and initially hesitated calling on his brother-in-law, James Dickson, because it was so early in the morning. After wandering around the streets in the vicinity of the British Museum, he found, by chance, that a door to the Museum was open and went in. To his surprise, Dickson happened to be there, tending to some matters; thus Park and Dickson were reunited, although it was unplanned and a rather inauspicious return.

Afterwards, Park was properly welcomed and received warmly, without the parades and hoopla that modern adventurers receive. Although Park enjoyed this acclaim for his exploits, he was disappointed in that he felt he did not accomplish all he wanted to do. He waited for another opportunity to travel. After his first journey, Park believed that the Niger River flowed into the Atlantic Ocean and was eager to test his hypothesis. It appears that this was the primary reason for his reluctance to accept Banks's offer to travel to Australia later on, not the reason generally given: that he did not want to leave his fiancé, Allison Anderson. His second trip to Africa makes the latter explanation suspect because when Park finally returned to Africa, he reluctantly left his family, a reasonably successful medical practice, and a feeling of general well-being. In 1803, another opportunity to return to Africa presented itself and this time it appeared to be properly financed and planned.

In June 1798, Park finally visited his widowed mother in Foulshiels, Scotland. He spent his time seeing his relatives and working on an account of his African trip and catching up with them for the time he spent away. His notes were somewhat

J. Schwartz, *Robert Brown and Mungo Park*, Memoirs of The New York
Botanical Garden 122, https://doi.org/10.1007/978-3-030-74859-3_8

disorganized due to his captivity and other travails. Initially, the material consisted of short rather disjointed notes written on separate pieces of paper. This imperfect journal contained considerable gaps, which he filled with information taken from his memory. He worked assiduously on his journal, with evening walks along the banks of the Yarrow River being his only means of recreation. Occasionally, he became more ambitious, exploring the rugged and wildly scenic areas nearby.

In the meantime, Park wrote to Banks to explain his position concerning his rejection to serve on the planned expedition to New Holland. In September [14, 1798] he wrote to Banks that he was "sorry that any misunderstanding Should have taken place respecting my present Voyage to new south wales—when you read me the Sketch of your intended letter to the Duke of portland I had every reason to be pleased with it—it was a plan of my own proposing and it was the hight [sic] of my ambition to have to have accomplished it with the honour." He explained that the wages "appeared too small and I should have mentioned it at the time … when the committee of the [African] Association should solicit for such a place things would be put upon a very different footing." Park, in effect, was telling Banks that his returning to Africa was a priority.[1] James Dickson wrote to Banks around the same time (September 20, 1798) explaining that he was "not in any way been a cause of Mr Park's not agreeing with your encouragement. … I have found out from his Sister which is my wife that thier [sic] is … a love affair in Scotland but no money in it…he is burying himself and his talents, a thing which both his Sister and me disapproves of much …."[2]

After a pleasant and busy summer, Park traveled to London with his manuscript.[3] He worked with Bryan Edwards (1743–1800), the secretary of the African Association (Association for Promoting the Discovery of the Interior Parts of Africa), who helped him edit his draft. He received further assistance from James Rennell (1742–1830), a leading geographer, skilled surveyor, and advisor to the African Association. Park was elected an honorary member of the Association and did editorial work for the Association, e.g., constructing maps of the areas he had explored. Rennell helped organize Park's notes and supplied a map of his own that showed the Niger flowing eastward ending in a trickle in a large swamp as Park postulated. He also provided additional illustrations and a route map for Park's publication (*Travels in the Interior Districts of Africa*).

The period of 1798-1804 proved to be quite eventful years for Park. His *Travels in the Interior Districts of Africa* was published, and was so successful that three editions were printed in English in addition to published French and German editions. Johann Friedrich Blumenbach (1752–1841), an anatomist and anthropologist

---

[1] Letter from Park to Banks, Sept. [14] [1798], letter # 348 in Sir Joseph Banks, *The Indian & Pacific Correspondence of Sir Joseph Banks*, ed. Neil Chambers, vol. 4 (London: Pickering & Chatto, 2011), pp. 548–549.

[2] Letter from Dickson to Banks, September 20, 1798, letter # 349 in *The Indian & Pacific Correspondence of Sir Joseph Banks*, p. 549.

[3] A posthumously published later edition of Park's *Travels in the Interior of Africa* reported that he quit "Fowlshiels with great regret," p. 338, *Travels* (Wordsworth Classics of World Literature, 2002).

and a founder of the discipline of scientific anthropology, wrote to Banks on September 19, 1798, exclaiming, "ardently I long to see once Mr. Park's own extensive Account of his wonderful & highly interesting Travels."[4] Although Park personally found the paganism in the tribes he had encountered repellent, and additionally was offended by the intolerance of certain practices of Islam he observed and experienced firsthand, his published account of his journey tried to put these thoughts aside. He provided an objective as well as a dispassionate narrative, including a glossary of the Mandinka vocabulary. Rennell's commentary on the significance of Park's geographical discoveries gave additional weight to the work.

Park's professional success was enhanced by his personal happiness. Park married his fiancé, Allison Anderson, the eldest daughter of Dr. Thomas Anderson, the man to whom he was apprenticed, on August 2, 1799. The Andersons were a second family to him, so he looked forward to a successful career in a beautiful region of Scotland, surrounded by family and friends.

Life as a Scottish Country Doctor

For the first two years of his marriage, Park and his wife lived with his mother and one of his brothers, who helped run the family farm. Then an opportunity to set up a medical practice in Peebles opened up. In October 1801, he became a Scottish country doctor on his own. For someone who had experienced exciting adventures and distant lands, he did not find this quiet life entirely satisfactory but there is little evidence that Park suffered any dissatisfaction, at least not at the outset. There have been suggestions that the drudgery of work as a Scottish country doctor with little reward or excitement began to wear on Park. Personally, he enjoyed domestic life and life in his home in Peebles, with his wife and growing family and circle of friends. Yet his romantic streak and his desire to finish the work he began on his first journey drew him to undertake further African adventures. He may have expressed certain misgivings about the drudgery of his medical practice to friends like Walter Scott: i.e., traveling on lonely Scottish roads in order to earn a meagre living.

In 1800, Gorée was captured by the British, prompting Park to write to Banks on July 31, 1800, explaining that this was an opportune time to again explore the interior of Africa, and as a consequence, he offered his services to be of use to his country.[5] The fall of Gorée made peace more likely between the French and the British, and a year later, in the autumn of 1801, Britain signed the "Preliminary

---

[4] Letter from Johann Friedrich Blumenbach to Banks on September 19, 1798, Letter #1484 in Sir Joseph Banks, *The Scientific Correspondence of Sir Joseph Banks*, ed. Neil Chambers (London: Pickering & Chatto, 2011), p. 554.

[5] Letter from Park to Banks, July 31, 1800. Various published biographies of Park, and edited additions to Park's *Travels*, contain passages from this letter but there is no indication where copies of this letter (containing the entire letter) are located. It is not included in The Banks Letters; A *Calendar of the Correspondence of Sir Joseph Banks, preserved in the British Museum, British*

Articles of Peace" with France. Now the African Association put aside their fears about undertaking dangerous expeditions. Banks wrote to Park, in October 1801, "that in consequence of the Peace, the Association would certainly revive their project of sending a mission to Africa; in order to penetrate to, and navigate, the Niger"; and Banks added, "that in case Government should enter in the plan, he [Park] would certainly be recommended as the person to be employed for carrying it into execution."[6]

When Park returned to Scotland, he gave careful thought about farming, but the high costs involved, i.e., securing good land, cattle, etc., made that proposition too risky in his view. Because an eminent surgeon in Peebles had passed away, there was an opportunity to succeed him but Park continued to think of returning to Africa. Park replied to Banks (on October 13, 1801) from Peebles informing him that he left London downhearted because his dream of settling in New Holland in a "romantic village" that he had imagined for himself and his family had been punctured. Park's idea of emigrating to Australia was a bit paradoxical considering his refusal to go to Australia as a naturalist. He indicated that his practice was entirely satisfactory but if anything presented itself—in view of peace with France, the "Association" would be again interested in reviving their mission to Africa to penetrate and navigate the Niger—he could hand over his practice to his younger brother (Alexander Park) or his brother-in-law (Alexander Anderson), telling Banks that he would "gladly hang up his lancet and plaister-ladle." He emphasized to Banks that he "always looked upon you as my particular friend," hoping that there were no longer any hard feelings concerning his rejection of Banks's offer.[7]

Obviously, Park felt confident enough that Banks was no longer angry with him so he felt free to write to him. However, again contrary to his expressed eagerness to go on another expedition in search of the Niger, he also harbored hopes of succeeding Dr. Daniel Rutherford (1749–1819) as chair of botany in Edinburgh and believed that Banks could help him in securing this position.[8] Obviously, Park was

---

Natural History Museum and other collections in Great Britain, Warren R. Dawson ed. (London: British Natural History Museum, 1958).

[6] Letter from Banks to Park of October 1801 indicating that "because of the peace, the Association would certainly revive their project of sending the mission to Africa, in order to penetrate to and navigate the Niger," p. 340 of *Travels*. It is not included in Warren Dawson's *Calendar* but the original is in the Selkirk Public Library. The Anderson family's collection of papers was divided up and scattered.

[7] Copy of this letter from Park to Sir Joseph Banks of Oct. 13, 1801, is in the Dawson Turner Collection, Department of Botany, British Museum (Natural History) *D.T.C.* 12. 265–266, and is in *The Indian & Pacific Correspondence*, letter # 6, vol. 6, pp. 9–10. Although Park turned down the opportunity to go on an exploring expedition to Australia, a few years later, he harbored thoughts of settling there with his family. Kenneth Lupton suggests that Banks may have been responsible for shattering Park's dream of settling in Australia, explaining that there were no "romantic villages" in Australia, only settlements for convicts, *Mungo Park, The African Traveler*, p. 125.

[8] Rutherford had achieved a great deal of fame because of his discovery of nitrogen. Coincidentally, when Rutherford eventually passed away in 1819, Robert Brown, then Banks's librarian, was offered the post but he refused.

quite ambivalent about his future and was concerned about earning a living for himself and his family. His wish to replace Rutherford indicates that he maintained a strong interest in pursuing a career in the sciences. It seems that his ambition to gain appointment as chair of botany in Edinburgh supports the notion that such a career path may have transcended his interest in conducting further exploration or continuing his work as a Scottish country doctor.

For the next two years, nothing further happened. While he was in Peebles, Park tended to his medical practice, settling into the life of a county doctor, and making friends in the community, including Walter Scott. Scott lived in Ashestiel, near Fowlshiels. He sought out Park and a firm friendship developed. There are contradictory accounts about Park's life as a country doctor. Some accounts indicate that he had adapted to his life as a physician in rural Scotland. Others have indicated that his three years "of a country practitioner" were spent "not contentedly."[9] But Park was dedicated and paid "willing attention to the poorest without thought of a fee." Scott supposedly said about the life of a Scottish country doctor, "there is no creature in Scotland that works harder and is more poorly requited than the country doctor, unless, perhaps, it may be his horse."[10]

Park continued to explore the rugged countryside near his home, and do some botanizing. He maintained his fascination with cryptogams (mainly mosses), an interest initially stimulated by his brother-in-law, James Dickson. His interest in mosses actually proved to sustain him at his lowest point when he was alone during his first journey, hungry and stripped of his possessions. The sight of the "extraordinary beauty of a small moss in fructification" lifted his spirits, giving him the resolve to continue writing his book, *Travels*. He received a great deal of satisfaction from his family. Happily married and now the father of three children, he enjoyed the friendship of his circle of friends—in addition to Scott—and his extended family including in-laws, the Andersons.

There was certain compensation to the hard work in Peebles in addition to his happy home life. He was friendly with Dr. Adam Ferguson who was a retired Professor of Moral Philosophy at Edinburgh University—where he had retired before Mungo became a student—and through Ferguson he met and became friendly with Ferguson's successor, Dugald Stewart. He became friendly with some of the privileged local society, such as Sir John Hay, Sir James Montgomery, and Colonel John Murray, who often invited him to dine. Montgomery convinced him to join the

---

[9] Stephen Gwynn indicated that "it was a laborious life and not without danger," because he traveled up to "remote glens where no wheel tracks existed," and "was at every patient's beck and call." *Mungo Park and the Quest of the Niger* (London: John Lane, The Bodley Head Limited, 1934), p. 149.

[10] Mungo Park. *The Life and Travels of Mungo Park, with the account of his death from the Journal of Isaaco, the substance of the later discoveries relative to his lamented fate, and the termination of the Niger.* (New York: Harper and Brothers, 1840), p. 158.

Yeomanry, the Home Guard that was formed several years before when there was anxiety about a French invasion.[11]

Did Park's relationship with and indeed sponsorship by the African Association indicate that he was sympathetic with the slave trade? In spite of his harsh treatment, he revealed little overt racism in *Travels*, particularly for that period of history. He demonstrated little interest in acting as a missionary, i.e., to convert the indigenous people he came in contact with to Christianity, although suspicion among some of the native people that he was interested in proselytizing may have been a negative factor in the treatment he received during his first trip to Africa.

On the other hand, his relationship with Bryan Edwards, a former planter from Jamaica and secretary of the African Association, could be interpreted as evidence that he was not against the slave trade. Edwards used his position in the House of Commons to oppose abolition of the slave trade. It could be argued that Edwards benefited from his relationship with Park, using Park's fame to his political advantage. While Park was by no means an abolitionist, perhaps he could be best described as an agnostic on the subject. His primary interest was to further the cause of exploration, geographical as well as in natural history, although his experience during the first expedition to Africa did not afford him much opportunity for discoveries in natural history. He certainly was drawn to adventure and the excitement of discovery of unexplored territory. No matter what his current circumstances were during the years 1801–1803, he waited for another opportunity to finish what he had begun in his first journey to Africa.

Park's Opportunity for Further Adventure

When Park received a letter from the Colonial Secretary State's office in the autumn of 1803 asking him to come to London for a meeting "without delay," he did not hesitate. He traveled to London where he had an interview with Lord Hobart, Secretary of State for the Colonial Department—he later became the Earl of Buckinghamshire—who informed him about a planned exploration to Africa and that he (Park) would play a major role, i.e., to organize an expedition to penetrate Africa's interior and navigate the Niger. There would be a guard of soldiers to be taken from the English garrison at Gorée—recently taken by the British forces—and Park would be given the rank of Captain. Park apparently suppressed his excitement with this offer, asking for a short period of time to think about the proposal and be able to discuss it with his family and friends. Prior to meeting with Lord Hobart, he wrote to Banks (on October 10th) informing him that he was contacted by Lord

---

[11] Mungo's extroverted brother, Archibald, was also a volunteer and Scott likewise in Edinburgh, *Mungo Park, The Africa Traveler*, Kenneth Lupton (Oxford University Press, 1979), p. 129.

Hobart, and speculated that he would be offered to lead another expedition and that it would "be set on foot."[12]

After his meeting with Lord Hobart, he wrote to Banks on October 20, 1803, that as he anticipated, an offer was "made to him …" to undertake "another expedition into Africa to discover the mouth of the Niger," and "would be allowed a guard of 25 soldiers, 10s. a day for subsistence and £200 a year while in Africa," adding that before he decided, he had to consult his wife and was returning to Scotland for that purpose.[13] Banks promptly responded that although he was "not sufficiently informed of the details of the proposed expedition to write on the subject" … he thought that offer was fair, and because "the remuneration offered seems handsome, he strongly advises acceptance of the offer."[14]

With Banks's encouragement, Park discussed it with his wife (Allison), and she did not put up serious opposition although probably she was not too happy with the prospect of Mungo leaving on potentially such a hazardous journey. She probably was additionally concerned when her brother, Alexander (Anderson), was asked to join the expedition. Alexander Anderson did not have the stamina that his brother-in-law Mungo had but was quite eager to go. Mungo helped him secure the position of surgeon in the expedition with the brevet rank of lieutenant.

Park traveled to London to prepare for the journey. He wrote to Alexander Anderson (on January 5th, 1804), "I arrived here [London] last night, and this morning had an interview with Sir Joseph [Banks] and Mr. [John] Sullivan [M.P. and Lord Hobart's Under Secretary] and find it is the wish of Government to open a communication from the Gambia to the Niger for the purpose of trade, and they do not wish me on any account to proceed further than I shall find it safe and proper …. I mentioned you as my companion, and that you expressed a wish to accompany me. They readily offered you 10s a day, and the Government allowance, with a promise of their interest in your favour."[15]

The following months were quite frustrating because the British Government already seemed to be ambivalent about launching a major military operation in the region. Colonel Stevenson had planned for a major military expedition from the Gambia to the Niger, and then to travel downstream to Timbuktu and thereby control the gold trade of the region, which would be mined, and bring an abundance of riches. Stevenson planned to use 1200 West Indian troops. The government was not completely supportive of Colonel Stevenson's plans. Additionally, the government at the time was not particularly strong, evidenced by its general paralysis; that is, they did not seem capable of making a clear decision. The expedition initiated by the Colonial Department (and supported by Banks) on the surface seems to be somewhat similar to the major military campaign that Colonel Stevenson had

---

[12] Letter from Banks to Park, October 10, 1803, *D.T.C.* 14: 161.

[13] Letter from Park to Banks, written from Stilton on October 20, 1803, *D.T.C.* 14: 162.

[14] Letter from Banks to Park, written from Revesby Abbey on October 26, 1803, *D.T.C.* 14. 163.

[15] Letter from Park to Alexander Anderson on January 5th, 1804, preserved in the National Library of Scotland. Copy of the letter in Kenneth Lupton's *Mungo Park, The African Traveler* (Oxford: Oxford University Press, 1979), p. 136.

planned but actually the commitment was rather modest. As a result of the government's indecision, Park and Anderson were in suspension, waiting around since the beginning of the year, but eager to begin their adventure.

Weary of the government's indecision, Park returned home to Scotland in March (1804). He brought Sidi Ombark Bouby with him, a Moroccan who came to London to serve as an interpreter to the Egyptian Ambassador. Park hired Ombark Bouby to teach him Arabic, believing that his problems on his first journey stemmed from differences in language. Park did not fully understand that cultural differences as well as often not understanding Park's motives and his position on the part of some native tribes were also factors in contributing to the misfortunes he had suffered during his first African journey. Ombark Bouby caused a minor sensation in Peebles, insisting on meat only from animals that were slaughtered according to Muslim custom, and avoiding any alcoholic beverages. He once grew very angry with Park for serving him a pudding that was prepared with brandy.

In May, Park gave up his home in Peebles and his practice, and he and his family (including Ombark Bouby) moved in with his mother and younger brother in Fowlshiels. Walter Scott, who began to achieve fame as a poet, rented a house nearby in Ashestiel. Scott had sought Park out previously, and the friendship between naturalist-explorer Park and Scotland's great writer grew. Park spent more time with Scott prior to his departure. The last time he saw Scott, Scott accompanied him part of the way home to Fowlshiels. When Park's horse stumbled, Scott said, "Ah! Mungo …. 'I am afraid that is a bad omen.'" The story succinctly conveys the forebodings shared by Park's family and friends about the journey he was about to embark on. Park replied, "*Freits* follow them that *freits* follow." This was a popular biblical expression in Scotland at the time.[16] There were reports that Scott literally begged Park not to go.

Banks had a major role in planning the journey. He wrote to the Secretary of State and Colonies in the latter part of the year (December 1804), submitting a "Particular Account" of the expedition, and sent a copy to Park. The stated purpose was similar to the plan outlined by Colonel Stevenson earlier in the year without quite the ambitious military commitment, i.e., "the expansion of British commerce and the enlargement of geographical knowledge, to examine the routes by which goods can be carried to the Niger, to record the physical features of the country, water supply, means of obtaining provisions etc."[17] The drastic cut in military forces proved to be a fatal mistake because the military contingent was designed to defend Park and his party and the decreased manpower was not up to the task owing to the risks of such an undertaking. Obviously, the foot-dragging and the reluctance on the part of the government framed these decisions, as well as the desire to protect Park, remembering what happened to him on his prior venture, but they ignored the reality that circumstances had changed. The presence of a small military contingent did

[16] *The Life and Travels of Mungo Park and an account of the progress of African Discovery* (Robert Chambers probable author) (Edinburgh: William and Robert Chambers, 1842). The loose meaning is "*freits* (omens) follow those who follow them (*freits*)."

[17] December 1804 letter from Banks to Lord Hobart with copy to Park. *D.T.C.* 15. 232–240.

not serve as a deterrent and may have been counterproductive because the sight of soldiers perhaps stirred up resentment and was inadequate at any rate. Park perhaps would have been better off conducting his explorations much as he had done before, or if he was to be "protected," it should have been a force more than adequate to allow him to explore and collect.

These issues were not so obvious in 1804 and Park waited throughout much of the year for the anticipated summons from the Secretary of State. In September, he received a letter from the Under Secretary of State for the Colonial Department, requesting that he should travel to London and present himself at the Colonial Office, and furnish Lord Camden with a written statement from him regarding the purpose of the mission, the means of carrying out, as well as the supplies and man-power necessary to achieve a successful outcome. He delivered a "Memoir" to Lord Camden on October 4, 1804. The overall object was "the extension of British Commerce, and the enlargement of our Geographical Knowledge." One of the things indicated here was that Park would "turn his attention to the general fertility of the country, whether any part of it might be useful to Britain for colonization, and whether any objects of Natural History, with which the natives are at present unac-quainted, might be useful to Britain as a commercial nation."[18]

The plan called for 30 European soldiers and six European carpenters as well as 15–20 "Negroes from Gorée," most of them artificers, i.e., skilled artisans or mechanics often selected from the army. Park also listed the equipment required, and an itinerary. The party would arrive at St. Jago, and there they would purchase donkeys and mules, as well as sufficient corn to maintain them during the voyage to Gorée and up the Gambia. At Gorée, they would take on the soldiers and artificers, and then go 500 miles up the Gambia to Fattatenda. With the permission of the King of Wooli, they would disembark, rest briefly, and then go on to the Niger, traveling through the kingdoms of Bondou, Kajaaga, Fooladoo, and Bambara. Park had been advised to renew his acquaintance with the King of Bambara—who did not see him during his previous expedition but gave him 5000 cowries instead—when he and his party arrived at the Niger. Boats were to be constructed in order to navigate down the Niger. Park concluded his "Memoir" (in effect, a travel prospectus) to Lord Camden, indicating that the mission was "second only to the discovery of the Cape of Good Hope," and geographically, "the greatest discovery that remains to be made in this world."

After Park delivered his "Memoir" at the Colonial Office, he had an audience with Lord Camden, who agreed generally with the plan but advised him to consult with Major Rennell—who previously had helped him prepare his *Travels* for publi-cation, Park's much celebrated account of his first journey. Rennell, who was fond of Park, was not convinced by the plan developed at the Colonial office. Rennell

---

[18] Excerpt from *The Journal of a Mission to the Interior of Africa in the Year 1805, Together with Other Documents, Official and Private, Relating to the Same Mission, to Which is Prefixed an Account of the Life of Mr. Park*, edited with commentary by John Whishaw (Adelaide, Australia: The University of Adelaide Press, 2012). The book was converted from a published account from the nineteenth century to a digital format in 2012.

thought that the proposed journey was too hazardous and the plans were too general and vague. Initially, Park was convinced by Rennell and was ready to abandon the mission but after some reassurance from other officials, these fears were allayed. But Rennell and Park's other friends continued to have serious misgivings.

Park's Fateful Journey

On January 2, 1805, Park received his instructions from Lord Camden, much along the lines of the "Memoir" he submitted to Lord Camden several months before. He was commissioned as a captain and was provided with a stipend of £5000 while Alexander Anderson was given a salary of £1000. For better or worse, he was ready to proceed. He wrote to his wife on January 8, 1805, his last letter to her before he left England, explaining his departure was at hand. He explained that he "got every thing I wished, yesterday I was made a Captain and Sandy [Anderson] a lieutenant, we are not obliged to wear any uniforms but have these Commissions to keep the soldiers in proper order."[19] On January 28th, he traveled down to Portsmouth with Alexander Anderson and George Scott—another friend from Selkirk whom Park chose to go as draftsman—and they boarded the *Crescent*. Two days later (on January 30, 1805), the *Crescent* set sail for Africa.

They arrived at Port Praya Bay, St. Jago, in Cape Verde Island (now called Santiago), on March 8th, passing by Fogo Island with its large volcanic formation dominating the entire landscape, described by Park as "the most sublime object I have ever seen." They were already behind schedule, because under normal conditions, the trip should have taken about two weeks but waves of stormy weather with immense amounts of rain and tides delayed their arrival in Cape Verde. In St. Jago, they took on 44 donkeys and large amounts of corn and hay although the Captain of the *Crescent* was not entirely happy with the ship taking on such additional cargo. Park wrote to James Dickson (March 18th) that his party was in "good health except [Alexander] Anderson" who was ill with rheumatism, and indicated that he "bought all the corn" needed and 24 donkeys and will buy 32 more (it turned out to be 20 more because the Captain did not allow more).[20]

With these provisions and rest, they set sail for Gorée (formerly the French island colony off the coast of Senegal, recently captured by the British). They arrived on March 25th, arriving at the port that still served as a slave center. When they disembarked there (in Gorée), Mungo presented his credentials to Major Lloyd, the Commander of the Garrison. He set about recruiting because he needed soldiers to fortify the party, and found many willing volunteers because rank and file enlisted men were quite bored by African service and the idea of adventure in the interior of the African continent appealed to them. They were also attracted by the promise of

---

[19] A copy of the letter in Lupton's *Mungo Park*, p. 151.

[20] *Dawson MS*. 46.28.

double pay and discharge from the Army when the journey was over. Consequently, Park was able to get the best men possible, choosing 35 privates led by an artillery officer, Lieutenant Martyn. Also, several sailors and four carpenters joined the group. Shortly before they left (on April 4th), Park wrote to his wife Allison that he was in good health and Alexander had recovered from the rheumatism he suffered as a result of the trip from St. Jago to Gorée. He reported that they took a 12-mile ride into the country. There was no mention if he did any collecting of plant or animal specimens. On April 6th, with some flourish, the party set out for the Gambia as the *Crescent* turned south (Fig. 8.1).

They arrived three days later at Jillifree on the Gambia where Park had last been 10 years before, a much younger and inexperienced man and quite alone then as well. With more responsibilities this time, he wrote out a report concerning the activities to be conducted to the government. There, they glimpsed an American ship headed for America and laden with slaves, and this grisly scene was made even more unbearable when the captain suppressed a slave revolt by shooting 14 of the slaves.

After this short visit, the *Crescent* sailed up the Gambia to the port of Kayee. They arrived in Kayee, a small town situated on the Gambia, toward the end of April, and there, on April 26th, Park wrote to Banks as well as his wife. He informed Banks that he had little time for natural history but was sending Banks three specimens including the *Fang jani* or self-burning tree of Gambia which grew in plentiful amounts along the "banks of the Gambia between Yanimaroo and Karee, and no where else." This unusual plant burnt by "some internal process," in Park's words,

ISLAND OF GOREE

Fig. 8.1 Island of Gorée where Park and his party landed at the start of his second journey to Africa, March 1805, from an engraving in Joseph Corry's *Observation upon the Windward Coast of Africa* (1807)

although he could not explain this implausible phenomenon. He also sent the *Kino*, the branch and fruit of the original kino tree and a paper of the gum, and the *Tribo*, a root with which the natives used to dye their leather a yellow color. He expressed his regrets that he was unable to send *shea* butter because of local wars in Bondou and Kasson and was unable to get any Bondou *Frankincense*. He also reported that he was in good health and sent his regards to Major Rennell.[21]

Park wrote to his wife the same day (April 26) that he was "uneasy" with the prospect of not receiving any letters from her until he returned to England. He reported that he was busy making preparations for the journey and that she might not receive any letters from him for some months but was in excellent health. He told her that the King of Kataba—the most powerful king in Gambia—visited the *Crescent* on April 20th and 21st and the King gave them, in effect, a safe conduct pass to the King of Wooli. He also wrote to James Dickson—also the same day— that if things went well, he expected to "drink all your healths in the water of the Niger."[22]

With this optimistic frame of mind and in generally good health, Park and his party departed from Kayee on April 27, 1805, to go to Pisania—where he had set out for the interior of Africa during his first trip almost 10 years before—to take on more provisions and donkeys as well as engage the services of Isaaco, a Mandingo priest. Isaaco was also a merchant and had experience in traveling inland so he was to serve as guide at this point. Actually, there were warning signs that belied his earlier optimism. The quality of his troops turned out to be second-rate, and because of the earlier delays, he would have to undertake part of his ambitious journey during the rainy season. The party not only would have to endure the debilitating heat but also faced the prospect of tornadoes and hurricanes that preceded and followed the rainy season. So Park had a critical decision to make. Should he postpone continuing the expedition until the end of the rainy season or should he try to reach the Niger before the onset of the rainy season? He decided to pursue the latter course, not fully aware of the dangers inherent with the decision he made.

After he finished his preparations, the party returned to Kayee briefly and then departed the town on May 4th for good, heading for Madina, the capital of Wooli, arriving there on the 11th of May. Before their arrival there, on May 8th, two soldiers came down with dysentery, always a danger on these types of journeys—the same type of problems suffered by Flinders and his crew on the *Investigator* but perhaps exacerbated by even harsher conditions. On May 15th, they arrived on the banks of the Gambia, and before they reached their next destination, Badoo, on May 28th, they were beset by an attack of a very large swarm of bees. Several of the donkeys died as a result, and the expedition was in danger of being abandoned from the confusion resulting from this setback.

---

[21] Park's letter to Banks, April 26, 1805, *D.T.C.* 15: 356–357.

[22] Letter from Park to wife, Allison Park, April 26, 1805, letter from Park to James Dickson, April 26, 1805. Letters in *Travels*, pp. 347–349.

In Badoo, Park took advantage of his last contact with the "outside" world by writing letters to Banks and to his wife, because they could be transported from the Gambia River to Britain. His letter to Banks, written on May 28th from Badoo near Tambacunda, explained that the "Slatee" he hired would allow him "to stop his asses till I write a few lines." He indicated that the expedition was "prosperous" so far and that they were proceeding on to Tambacunda. He reported that he had observed the latitude every "two or three days" and observed three eclipses of Jupiter's moon (referred to as satellites), noting the longitude, and also observed many *shea* trees. He mentioned that the "course of the Gambia" was charted too much to the South but had determined "nearly its whole course." He wrote that he expected to reach the Niger on June 27th. Quite poignantly, he reported that he was "still alive."[23]

Park's letter to his wife written hurriedly from Badoo a day later (dated May 29, 1805) was more optimistic sounding. Park told his wife that they were "half through" their journey "without the smallest accident or unpleasant circumstance." He told her that by the end of June, he expected to have finished "all our travels by land" and when they "got afloat on the river, we should conclude that we are embarking for England."[24] This false bravado masked the perilous shape the expedition was in. The rainstorms from tornadoes (hurricanes) increased in intensity, and this not only hindered their progress but also endangered the health of nearly everyone in the party. By June 10th, the rains were constant and 12 of the men became very ill. And they still were only halfway to the Niger.

After arriving at Shrondo (in the Kingdom of Dentila), conditions worsened although Park noted the gold mines in the area and his account in *Travels* devoted considerable space to how the mineral was collected and a description where it was located. On June 12th, they were hit by a ferocious tornado that forced them into the huts of the natives, storing their gear there as well. The party trudged on to Dindikoo, and Park was quite impressed with the beauty of the mountainous area and how well it was developed (and the happy state of the natives). Even so, their situation did not improve. They changed their course and did not return to the Gambia but instead proceeded on a northeast direction, thereby avoiding the Jallonka Wilderness. Travel became even more perilous, and disease afflicted many members of the party, exacerbated by the continual rain.

At the beginning of July, they almost lost their guide, Isaaco, from a crocodile attack, and by July 6th, most members of the party were either sick or were in greatly diminished circumstances. Still, they moved on, with only 11 European survivors out of the 40 when they left the Gambia, including Park, Anderson, Scott, and Martyn, but both Anderson and Scott became seriously ill. Scott had to be left behind in Koomikooki as they continued to travel to the Niger, and he died shortly after before he could see the Niger. Park remained in acceptable shape as he tried

---

[23] Letter from Park to Banks, May 28, 1805, *D.T.C.* 16: 39–40. Also in *Journal of a Mission*, ed. John Whishaw, p. 41.

[24] Letter from Park to Allison Park, May 29, 1805, in *Journal of a Mission*, pp. 41–42. Also in *Travels*, p. 351.

valiantly to hold things together. The party reached the Niger, at Bambakoo, on August 19th, and they proceeded down the river on the 21st. In the meantime, he sent Isaaco to Sego, the capital of the Kingdom of Bambara to negotiate free passage with Mansong the King so they may accomplish passage into the interior. While waiting for Isaaco's return at Marraboo, Park came down with dysentery but he was able to knock it out of his system with a strict regime of medicine (mercury), and his strong constitution helped rid himself of the malady.

Negotiations with Mansong's ministers were as difficult as they were during Park's first journey when he arrived at Sego. At that time, Mansong refused to see him and demanded that he should immediately leave Sego. Mansong remained mistrustful and was fearful of the reaction of his Moorish subjects. However, Park was allowed to go to Samee, near Sego, and then to Sansanding where he could build a craft for the voyage down the Niger. Revised editions of *Travels* published after Park's death bear witness to the daily calamities that beset the party; "Aug 9th— Michael May, having died during the night, buried him at daybreak …. Hired people to carry down bundles and put them into canoe … that the soldiers had nothing to move, being all weak and sickly."[25] "Aug. 10th—William Ashton declared that he was unable to travel; but as there was no place to leave him at, I advised him to make an exertion and come on …."[26] "Aug. 11th—This morning hired Isaaco's people to go back and bring back and bring up the loads of soldiers who had halted by the side of the stream. In the course of the day all the loads had arrived … in … the last two marches we had lost four men, viz., Cox, Cahill, Bird, and Ashton." He also recorded that "Mr. Anderson still in a very dangerous way, being unable to walk or sit upright." Subsequently, they tried to put Anderson on a horse because it was unrealistic to leave him.[27]

On August 14th Park reported that Jonas Watkins had died, and Anderson was carried in a hammock. The donkeys (asses) were also ill and the men were afraid to allow them to eat the corn of the natives, because the law of Africa determined if the animals eat the corn, then the owner of the corn takes possession of the animals. The situation continued to deteriorate during the remainder of the month. In addition, Park was concerned about Mansong's coldness, which Park assumed was caused in part by religious intolerance from the Moorish subjects of Mansong, who were also apprehensive by the threat of commercial rivalry. This anxiety was relieved a bit by the arrival of Bookari, the "singing man or bard" of Mansong, on September 13th, who had six canoes which could facilitate their departure from Marraboo. The same day, he left Marraboo and went down the Niger but did not stop at Sego because of Mansong's hostility in spite of the gifts that Isaaco had presented to him on behalf of Park. Instead, he traveled to Sansanding and stopped there.

Sansanding was an area east of Sego on the Niger with 10,000 inhabitants. Park spent most of the next two months there, primarily to construct a proper craft

---

[25] *Travels*. p. 355.

[26] *Travels*. p. 356.

[27] *Travels*, pp. 357–358.

to continue navigation down the Niger. To do so, he constructed a flat-bottomed boat from two canoes that he was able to acquire in trading with the native population but they were in poor shape. Park sold his surplus supplies and he spent weeks working on the construction of the boat with his own hands, and when it proved to be suitable for travel down the river, he named this vessel, His Majesty's Schooner, the Joliba.

While in Sansanding, Park conducted little or no natural history investigations although he remained fascinated by the beauty of the area. He had little time for such research because he was busy negotiating with the local population. He also had to deal with more heartbreaking news; he got word that George Scott, whom he had to leave behind at Koomikoomi, had passed away, and most crushing of all, the devastating death of his good friend and brother-in-law, Alexander Anderson, on October 28th. This cruel blow still did not persuade him that this entire venture was futile despite the prospect of having to navigate an unexplored river in a vast continent full of danger and disease, and without any close companions except Lieutenant Martyn and only three soldiers—one in quite a perilous mental condition—remaining from the original group.

When he had finished construction of the schooner (*Joliba*), he sat down to write letters to his father-in-law, Thomas Anderson, Sir Joseph Banks, Lord Camden, and his wife, Alison. He reported to Banks that he had sent Lord Camden a full account of his daily movements and asked him to share these remarks with Banks. He repeated his determination that he would follow the river to its termination and that he secured a guide—who had traveled widely and knew a great deal of the country they were about to explore—to accompany him to Kasha. The guide told him that the river bended sharply southwards but did not know anyone who had seen its termination point. Finally, showing some optimism, he told Banks that he had purchased some shea nuts which he promised to take to the West Indies because he intended to return that way, and hoped to reach the coast in three months.[28]

Park wrote to Earl Camden the following day (November 17th aboard H. M. Schooner *Joliba*, "at anchor off Sansanding"). He mentioned that his trepidation about the rainy season and its effect on Europeans had proven him correct because the journey from the Gambia to the Niger had cost the lives of 39 men out of the original 44 Europeans who had begun this difficult trip. Only five were still alive, three soldiers, Lieutenant Martyn, and himself. Nevertheless, he was not despondent because he was determined to "discover the termination of the Niger or perish in the attempt" on the "tolerably good schooner" he had constructed, and with a bit of flourish, he hoisted the British flag. He added that if he succeeded, he expected "to be in England in the month of May or June by way of the West Indies."[29]

The next day, Park wrote to his wife and it must have been the most troubling effort he ever had to undertake. He wrote to her that it grieved him "to the heart" that

---

[28] Letter from Park to Banks, dated November 16, 1805, written from Sansanding. B.M. ADD. MS. 37,232. 64–65; *D.T.C.* 16: 159–160. Also copy in *Journal of a Mission*, pp. 46–47.

[29] Letter from Park to Lord Camden, Nov. 17th, 1805. Copy in *Journal of a Mission*, pp. 47–48.

her "brother Alexander, my dear friend is no more!" He related that "the greater part of the soldiers [as well as her brother and Scott] have died on the march during the rainy season," but he assured her that he was in good health. Now that the rainy season was over, he believed that there was still a "sufficient force to protect" him from "any insult in sailing down the river, to the sea." He added that he intended to set sail as soon as he finished the letter and not stop or land anywhere until they reached the coast. Then he would set sail for England.[30]

This was the last anyone heard of Mungo Park. With only Martyn and three soldiers remaining, they sailed down the river without stopping. They had adequate supplies on board but they failed to get permission from local chieftains whose territory they were traveling through. Park hoped to race toward the ocean but the Niger continued on south. Accounts of his last journey surfaced when Isaaco left Senegal in January of 1810, and after an absence of 20 months returned on September 1811, confirming the reports of Park's death. Another journal, compiled by Amadi Fatouma, the guide who accompanied Park from Sansanding down the Niger, corroborated Isaaco's testimony.[31]

Both accounts reported that Park and his party in 1806 were slowed down by rapids and then were attacked from shore near Bussa, similar to problems they had faced previously, but their response was quite foolhardy in that they returned fire, trying to shoot their way out of this precarious situation. They were unlucky in employing this strategy on this occasion: Park, Martyn, and the soldiers died, possibly from drowning. Park's death ended the exploration of the Niger until the conclusion of the Napoleonic Wars in 1815; its mouth was finally reached in 1830, in the Gulf of Guinea, where it forms one of the largest deltas on the Atlantic Ocean.[32] Park's life and career had come to a sorry end, and the British public and all Scotland mourned him as did his wife, three sons, and daughter, but the public's imagination was fired by his daring and determination.[33]

---

[30] Letter from Park to his wife, Allison Park, November 19, 1805, in *Travels* pp. 380–381. Also a copy in *Journal of a Mission*, pp. 48–49.

[31] The document was given to Colonel Maxwell and from him it reached the Secretary of State for the Colonial Department.

[32] The source of the Niger River is in the Fouta Djallon Mountains of Guinea near its border with Sierra Leone. It flows north-east and then turns south-east, before ending in the Gulf of Guinea, forming one of the largest deltas in the world.

[33] For a number of years, there were misleading reports that he was still alive. Nathaniel Pearce, a traveler and adventurer as well as a servant to Lord Valentia and Henry Salt, Valentia's secretary and draftsman, after running away to sea several times, became a botanist. He wrote a humorous account of his adventures in Abyssinia. Pearce stated without equivocation that Mungo Park was still alive in 1818 and "was in honourable captivity at Timbuctoo," page 284, f.n. 1 of Edward Smith's *The Life of Sir Joseph Banks, President of the Royal Society* (London: John Lane, The Bodley Head Society, 1911/New York: John Lane Company, 1911). Years later, his second son, Thomas, went to Africa to look for him or find out precisely the details of his last days. He perished of fever in this futile attempt, in 1827.

Park's brief but colorful and promising life (and premature passing) indicates that he should have been able to accomplish more had he lived. Because of the circumstances of his death and the rapid pace he took through largely unexplored territory, he had little opportunity for any significant work in natural history so the impact of his expedition was confined to what he accomplished in the area of geographical exploration. In retrospect, the tragic circumstances of Park's death (as well as his promise), indicates the foolishness of undertaking expeditions without proper preparation and without sufficient safeguards. It seems that his life was unnecessarily sacrificed and those in the government and perhaps even Banks himself did not do enough to protect the men when they were sent out on such hazardous journeys. The zeal and taste for adventure that drove such men as Park made them susceptible to those who treated them as expendable objects.

A genus of tropical trees, *Parkia*, was named in his honor.[34] His brother-in-law, James Dickson, lived for 16 years after Park's death, passing away at the age of 84. It was up to his contemporary, Robert Brown, to carry the torch for British botany and natural history for the next 50 years. He did so ably (Fig. 8.2).

Fig. 8.2 Depiction of the death of Mungo Park, from *Travels in the Interior of Africa* (Edinburgh: Adam and Charles Black, 1858)

---

[34] *Parkia* is a genus of leguminous flowering plants commonly referred to as nitta trees, bearing red or yellow flowers and edible seeds.

Chapter 9
A Tedious and Uncomfortable Passage

Contents

While the *Investigator* was exploring in the Gulf of Carpentaria (in November 1802), it showed unambiguous signs that it could no longer function properly. This judgment was confirmed as it sat in Sydney Harbor later on in June 1803.[1] It became clear to all parties that the ship would have to be declared unseaworthy, and the voyage would have to be shortened. Flinders now had to find another ship to return to England. Brown and Bauer were freed from the obligations of serving on the *Investigator* and were now left to fend for themselves.

Taking the *Investigator* out of service turned out to be a fortuitous circumstance for Brown. Its condemnation meant that he had a choice: either to return home with Flinders and wait with him as he obtained another ship or to remain in Australia and continue exploration. If Brown had decided to sail home then, the work he accomplished up to this point would have represented a considerable accomplishment in natural history but even so he was not satisfied. He felt that further work needed to be done and he seized upon this fresh opportunity to do so. Driving his desire to stay on was his apprehension over whether his collection of specimens would safely reach England. The colonial authorities were happy to allow him to remain and work with local colleagues. Brown and Bauer were given permission to stay on to explore, observe, and collect.

Brown had already collected nearly 4000 species of plants and a sizeable amount of zoological specimens as well, while Bauer had made numerous drawings. One

---

[1] William Scott of H.M. armed vessel *Porpoise*, Edward Hanmore Palmer, commander of the ship *Bridgewater*, and Thomas Moore, who was master boatbuilder in the territory, concluded that the *Investigator* was so defective that it was not worth repairing. Their report was transmitted on June 14, 1803. *Nature's Investigator*, p. 411.

© The Author(s), under exclusive license to Springer Nature Switzerland AG 2021   121
J. Schwartz, *Robert Brown and Mungo Park*, Memoirs of The New York
Botanical Garden 122, https://doi.org/10.1007/978-3-030-74859-3_9

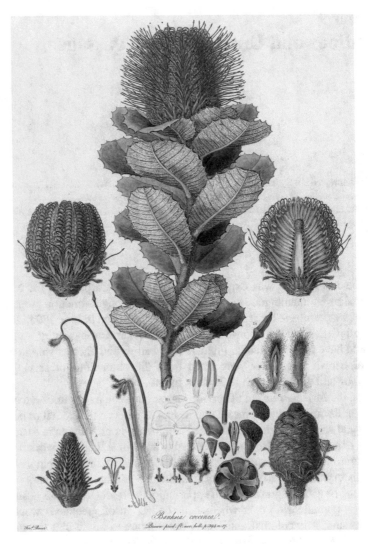

Fig. 9.1 *Banksia coccinea* (Scarlet Banksia), by Ferdinand Bauer, State Library Victoria, 30328102131488/4

such example of the exotic flora collected by Brown was the brilliantly colored species of plants native to Western Australia belonging to *Banksia coccinea* (meaning scarlet colored). The genus was named after Joseph Banks and is still commonly known as the scarlet banksia (Fig. 9.1).

In the meantime, the seeds that Brown and Good had collected reached England. In his April 10th letter to Flinders, Banks praised him for the "very gallant & successful manner in which you have executed … what I conceive to be the most difficult part of your Task the S. West part of the Land." He also expressed gratitude for "The Frequent opportunity you have given to the naturalists to investigate does you Great Credit both as a navigator & as a considerate man. Natural History is now a Study so much in Repute with the Public & in its self so interesting that the Good word of the Naturalists when you Come home will not fail to interest a large number of People in your Favour."[2] Banks emphasized his gratitude to Flinders for giving Brown the freedom to collect the seeds of so many different species. Banks again reflected that he wished Cook had given the same consideration to his naturalist, namely himself.

Despite the setbacks incurred in the expedition, Banks was pleased with the progress of the journey at this stage, having not yet read Brown's letter of March 1803 with its more pessimistic tone. He wrote to Brown (on April 8, 1803), reporting that he was in good health and "had been so fortunate in the first part of the interesting business in which you have so handsomely volunteered yourself." Banks was also full of praise for Flinders, "Your Commander deserves in my opinion great credit ...." and sent best wishes to Bauer for his "meritorious labours" which he called "indefatigable." He informed Brown that the "Seeds you sent by the Ship that brought your Letters came safe & in good order to hand they are all Sown in Kew Gardens & much hopes built on the success of them which we expect will create a new Epoch in the prosperity of that magnificent Establishment by the introduction of so large a number of New Plants as will certainly be obtained from them." Banks added a comment about Britain's relationship with other countries, "Tho' our Peace with France proves no more turbulent & quarrelsome Truce—Botany advances rapidly I am in hopes just at this time of sending a Gardener to China to reside there some years & send over to Kew Annual supplies of the beautiful & valuable products of that Country."[3]

This comment was very much in keeping with Banks's philosophy concerning international rivalries and the challenge of conducting scientific research during a period of conflict. Science was paramount, and scientists were to be treated with respect and were entitled to be secure. He also recognized how good relations between nations would help to foster not just international understanding but that there was much potential benefit for the countries involved as well, i.e., increased trade, economic gains, and advancement of science. Perhaps he was influenced by the example set by Captain Cook, under whom he served earlier as a young man. However, this attitude was not universally shared, as Flinders was to discover shortly afterwards when the French imprisoned him when hostilities between France and Britain broke out again.

---

[2] Letter from Banks to Flinders, April 10, 1803. Copy of original letter in *British Museum/Library*, Add. MSS 32439 f. 95. Copies in *D.T.C.* 14. 55–57, and in *The Indian and Pacific Correspondence of Banks,* vol. 6, Letter # 110, pp. 161–162.

[3] Letter from Banks to Brown, April 8, 1803, *British Library* Add. MS. 32439.95 & *D.T.C.* 14 43–45. The date of the letter listed in the Dawson Catalogue, April 8, 1803, although April 5th is given as the date of the letter in some places. Letter also in *The Indian and Pacific Correspondence of Banks,* vol. 6, Letter # 109, pp. 160–161.

While Flinders set out to find another ship—the possibilities were the *Rolla* or *Porpoise*—Brown resumed his fieldwork (on June 16th). Beforehand, he had to deal with the consequences of Peter Good's death from dysentery (which occurred on either June 11 or 12, 1803). Brown indicated in two letters to Banks—written from Sydney on August 6, 1803, and another with an undetermined date but written after September 1803—that he would, at Flinders's request, take charge of Good's post-humous affairs, supervising the sale of his possessions, and organize Good's papers and personal items. In an eloquent tribute to Good, Brown wrote to Banks, "Good, who while enjoy'd health, was most indefatigable, & whose curious exertions in his department were without doubt the cause of his untimely fate, died a few days after our arrival here [Sydney, New South Wales] of Dysentery contracted soon after our departure from Timor."[4]

Good's diary dutifully recorded events until several weeks before his death. His last entry, from May 17, 1803, noted, "At day break past Termination Island & about 8 AM past Bay No 1 keeping south of all the Islands—past noon departed this life Charles Douglas Boatswain of Dysentery with which he had laboured since the middle of Aprile—Self and several of the Crew labouring under the Same disorder."[5] Good was buried in a churchyard in Sydney on June 13th, and Brown and others from the *Investigator* attended the popular collector's funeral.

Several days afterwards (on the 16th), Brown explored the north shore of Port Jackson but both Bauer and he were still feeling the effects of the illness that had taken the lives of Good and other members of the crew. Flinders also had a setback; his hopes that the *Rolla* could be purchased to replace the *Investigator* were dashed when Governor King indicated that he was unwilling to incur additional expense. Instead, King suggested that the *Porpoise* might be available once it returned from its surveying.

Because Brown and his draftsman, Bauer, remained in a weakened condition, they were at first unable to take full advantage of the time they now had in Sydney. Flinders began to write a long letter to his wife, which he wrote in bits and starts but found himself unable to finish for more than a year. The letter is quite informative and useful today in that it contains his judgment about those who served under him. Flinders observed, "Mr. Brown is recovered from ill health and lameness." The draft of his letter included Flinders's curious remark that "we are not altogether cordial, but our mutual anxiety to forward far the complete success of the voyage is a bond

---

[4] Letter from Brown to Banks, August 6, 1803, British Library Add. MS 32439 ff. 104–108. Also copy in *The Indian and Pacific Correspondence of Banks,* vol. 6, Letter # 152, pp. 268–273. The loss of Good's commentary after his death has made it more difficult to decipher Brown's recorded notes in his diary as readily because Good was a very reliable diarist. Brown's sketchy descriptions of his work and observations are clarified by Good's reports. See *Nature's Investigator*, p. 409, for further discussion of the impact of Good's death in describing the history of the voyage. A subsequent letter from Brown to Banks [no date but also written from Sydney] informing Banks that Good's "effects have been sold," with the exception of his papers, *British Library Add. MS* 32439. ff. 100–101.

[5] *Journal of Peter Good*, p. 122.

of union; he is a man of abilities and knowledge, but wants feeling kindness."[6] Flinders's comment does not further elaborate on his judgment of Brown. His public remarks about Brown—in his letters to Banks for example—were always very positive. There is nothing in the published accounts of the expedition to indicate that there were any bad feelings between the Captain and the naturalist so the comments he made about Brown in his letter to his wife are somewhat inexplicable. Perhaps Flinders misinterpreted Brown's reserve as being unfeeling and cold. Flinders may have been bothered by Brown's behavior concerning Good's death. There seemed to be little feeling expressed by Brown about Good's passing, i.e., how much it was a serious blow and a personal loss. Brown was not the type of person who bared his emotions in this way. He went about his work, and dealt with arrangements for Good's funeral and the dispersal of the latter's possessions in a business-like manner so perhaps this is what bothered Flinders.

Brown continued his fieldwork in the Sydney area. After some negotiations, Flinders decided to return to England on the *Porpoise* under the command of Robert Fowler who was Lieutenant on the *Investigator*. William Scott, who was commander of the *Porpoise*, offered to relinquish command of the *Porpoise* to any of the officers of the *Investigator* but it was determined that Flinders, Westfall, and Allen would travel as passengers on the *Porpoise* after Flinders put the *Investigator* out of commission by arranging the discharge of the officers and other men who were on the *Investigator*. Governor King thought that this was the most prudent course of action.

Brown and Bauer were allowed to remain in Australia and continue their investigations. Brown and Bauer wrote to Flinders (on July 13th) that because Flinders planned to return to England immediately, they decided to stay rather than go home. The *Porpoise* was refitted to accommodate Flinders, and the greenhouse from the *Investigator* was moved to the *Porpoise* along with the botanical specimens collected by Brown and Good. On August 6th, in addition to his long letter to Banks, Brown wrote one to Sir Charles Francis Greville, Banks's close friend, a collector of antiquities, minerals, and other curiosities, as well as a politician. He was an important man to have on one's side, and Brown acknowledged this by keeping him informed.

Brown's letter to Banks summarized the journey up to that date, and expressed his deep regrets over Good's death, and his need for an assistant to replace Good. He expressed thanks that he had the opportunity of surveying the area where the ship landed but the collections in zoology were sparse and explained that mineralogy was "a barren field" and "even botany has fallen short of" his "expectations." He also mentioned the garden on the *Investigator*, and that it was not tended properly due to "the carelessness of the people on board," and it was much inferior to the one

---

[6] See Geoffrey C. Ingleton *Matthew Flinders Navigator and Chartmaker*, 2 vols. (Guildford, England: Genesis Publications Ltd., 1986). Ingleton indicates the Flinders began his letter to his wife Ann on 25 June 1803. Flinders to Ann Flinders, 25 June 1803, CY 1090 Safe 1/55. Private Letter Book, Vol. I, ML.

"that the Porpoise brought from England." He indicated to Banks that the garden from the *Investigator* should be put on board the *Porpoise* in order to transport the plants safely to England.[7]

Brown's Journeys from Sydney

Once their plans were set, Brown and Bauer spent their time collecting the better part of August in the Sydney area. The *Porpoise* sailed on August 10th with Flinders on board. Because Brown was already busy collecting and as he was feeling stronger, the pace of his work increased. His collecting was not just restricted to plants but he also collected zoological specimens. He continued his work in the Sydney area, studying *Hibbertia dentata* (commonly known as the toothed guinea flower), a native plant found on the east coast of Australia in the margins of rain forests, in shady areas and plants that were popular in hanging baskets; *Hibbertia virgata* (twiggy guinea flower), a shrub that can grow to 150 centimeters high, with stems containing crinkly hairs, and found in the Sydney area, particularly in the southern regions of the district; *Platylobium microphyllum* (small-leaved flat pea), another native plant of Australia; and *Pultenaea emarginata* (a bush pea), also native to the Sydney area. Brown remained in Sydney while Bauer worked in nearby Parramatta.

On August 17th (1803), Flinders suffered further misfortune. As the *Porpoise* sailed in the southern Coral Sea, it ran aground on the fittingly named Wreck Reef. Flinders reported the disaster in September and spent the time attempting to salvage what he could of the seeds and plants that were on board. On September 14, 1803, Brown wrote to Banks explaining that "the unfortunate loss of the Porpoise gives me an opportunity" to make up for the loss of the plants. Flinders reported that most of the seeds were safe although some were lost. Brown informed Banks that the four boxes stored in the greenhouse were successfully removed. Brown also informed Banks that he planned to visit Van Diemen's Land (Tasmania) and Norfolk Island (where the most notorious penal colony was located).

Brown repeated his plea to Banks that he needed an assistant: "I have procur'd some boxes for the garden brought out in the Porpoise, and have begun to establish in them such plants as I have reason to believe have not been sent to England; but in this department I stand in great need of a proper assistant, the person whom I have at present being so old that I cannot take him more than 2 or 3 miles from & so little of a Gardener that the greater part of this business devolves on myself." He also

---

[7] *HRNSW* (*Historical Records of New South Wales*) 5: 170–171. The two letters have been transcribed in *Nature's Investigator* from copies in *HRNSW*. 5: 180–187. Both of the original copies, dated August 6, 1803, are preserved in the *British Library* (*B.L.*) *Add. MS* 32439 ff. 52–53, 104–108. A copy of Brown's letter to Banks is in *The Indian and Pacific Correspondence of Banks*, vol. 6, Letter # 152, pp. 268–273.

reported collecting "a new and remarkable species of *Didelphis* (Opossums), likely *D. lemurina* (the Common Brush tail Possum) or *D. obesula* (the Short-nosed Bandicoot)."[8]

Tasmania

Although there were no plans to explore the territory of Tasmania (Van Diemen's Land) extensively when the itinerary for the trip was originally planned, Governor King was interested in making British settlements there. King entrusted Lieutenant Governor David Collins to travel to Port Dalrymple and to the Derwent River to determine the most favorable spot to settle because Collins was dissatisfied with conditions in Port Phillip. Brown was offered the chance to join the group at the end of 1803 and he seized upon this opportunity, traveling without Bauer who remained behind in Sydney. At the end of November (the 28th), Brown sailed on the *Lady Nelson* to Tasmania. In Bass Strait, there were heavy seas and squalls from gale force winds. This delayed his passage.

Brown wrote to Banks while they were anchored at the Kent's Group, Ross Straits (Dec. 30, 1803), "it is now upwards of a month since I left Port Jackson on the Lady Nelson, intending to examine the neighbourhood of Port Phillip & hoping also to have an opportunity of visiting Port Dalrymple & the Derwent in Van Dieman's Land." He referred to his collection, "consisting of *twelve Pucheons* containing specimens of Plants & *four boxes of Seeds* which last were sent back from *Wreck Reef.*" Brown concluded his letter by explaining to Banks that "in one of our attempts to reach Port Phillip from this anchorage we met the Calcutta at Sea, but I had not the good fortune to receive any letters or packages by her. I conclude they are under cover to Governor King."[9]

Because this was Brown's second visit to the Port Phillip area, having last visited in May 1802, he had the opportunity to try and make up for the loss of his previous collections despite his bravado about not being concerned about the misfortune he suffered by the loss of specimens gathered during his previous visit there. When he first visited Port Phillip (from December 1801 to May 1802), he had filled his time there observing and collecting. Brown considered the work he had done then as the

---

[8] Letter from Brown to Banks, September 14, 1803, *B.L. Add. MS.* 32,439, 131–132. Also a copy of Brown's letter to Banks is included in *The Indian and Pacific Correspondence of Banks,* vol. 6, Letter # 162, pp. 293–294.

[9] Letter from Brown to Banks on December 30, 1803, Robert Brown Letters, Natural History Museum, Botany Library, *D.T.C.* 14:180–181. Also copy included in *The Indian and Pacific Correspondence of Banks,* vol. 6, Letter # 174, pp. 306–308. Because hostilities had resumed between Britain and France, Brown understood that his work would have to be conducted under difficult circumstances. Brown's *Diary* was rather sparse and disorganized at this point, probably because he could no longer call on Peter Good's excellent skills as a diarist.

most productive he had accomplished up to that point, and now he relished the opportunity to expand on this previous work.[10]

The *Lady Nelson* finally reached Tasmania on New Year's Day of 1804 at Port Dalrymple (George Town). Brown was relieved to have reached Tasmania (Van Diemen's Land) after the terrible conditions he had endured during the passage there. Brown did a bit of collecting and discovered a new taxon of *Eucalyptus* (*Eucalyptus amygdalina* Labill), belonging to the Myrtaceae family. He noted that it was the largest of the trees he observed there. He also observed *Mimosa verticillata* (*Acacia verticillata*) during this brief visit. He also reported his first contact with aborigines.[11]

The *Lady Nelson* voyaged up the Tamar River on January 7th and Brown spent some time observing the geology in that region. Despite his initial negative impression of Tasmania, he wished to do further study there so he may have been somewhat disappointed when the *Lady Nelson* returned to Port Dalrymple, and sailed from Tasmania back to Port Phillip on January 21, 1804. Brown was anxious to continue exploration of the area near Port Dalrymple, the Tamar River and the Derwent River. The *Lady Nelson* returned to Tasmania, Risdon Cove on the River Derwent, with Brown on board on February 9th. His decision to remain in Tasmania to continue his explorations meant that he wound up staying there until August of 1804. Brown was eager to continue the work he had begun earlier, feeling that he had just scratched the surface of possibilities of discovery in natural history. Governor King and Lieutenant-Governor William Paterson supported Brown's decision to stay, mindful of Brown's powerful patron (Banks). When the *Lady Nelson* eventually returned to Sydney on March 4th, Brown was not a passenger on board for the return passage.

From January to August 1804, Brown busied himself making observations and collecting despite being handicapped by the sparse resources at his disposal. The primary purpose of the expedition from the government's perspective—but not Brown's—was to solidify colonization of the territory of Van Diemen's Land, not in conducting scientific investigation. However, Brown prudently took advantage of the availability of another vessel, the *Ocean* that was anchored at Risdon Cove in the Derwent River, using it as his base of operations.

The *Ocean* began its journey up the Risdon with Brown, mineralogist Adolarias William Henry Humphrey (1782–1829), and David Collins (1756–1810) on board. Humphrey had traveled from England in 1803 with David Collins to help establish a colony in Van Diemen's Land. Like Collins, Humphrey was dissatisfied with conditions in Port Phillip. He journeyed up to the Derwent River to search for

---

[10] *Nature's Investigator*, p. 209, note 13.

[11] *Nature's Investigator*, pp. 467–468. Brown's notes of the Tasmanian adventures in his Diary are rather sparse. Brown no longer had the able assistance of Good and even Flinders. They were written in light pencil and were in fragments. Moreover, his dates are slightly incorrect but fortunately his labeling of the plant specimens he collected are accurate. Also, because Bauer remained behind in Sydney, Brown did not have the benefits of Bauer's illustrations that previously enhanced his observations. *Nature's Investigator*, p. 459.

freshwater and minerals, rejoining Collins and Brown as they explored the Derwent, with Brown collecting important botanical specimens. Collins wrote to Banks from Hobart on July 20, 1804, informing him about his progress, i.e., clearing "a Garden" which would be ready for the "Plants & Seeds" he anticipated receiving. Collins wrote to Banks that he "met at Port Philip & ... accompanied ... that sensible & scientific Traveller, Mr. Brown. He has made several Excursions round this Neighbourhood & I understood has enriched his botanical Collection with some rare & curious Plants. He has very obligingly favoured me with a List of Trees and Shrubs, which he found growing in this Vicinity, a Copy of which presuming it may not be unacceptable."[12]

Brown, obviously feeling the loss of Peter Good and the support he enjoyed earlier, was not entirely happy with his situation but remained resolute nevertheless, determined to make the best of the opportunity to explore the territory and supplement what had been saved from the losses suffered when the *Porpoise* had been grounded. In spite of the reservations he expressed in the letter he sent to Banks at the end of 1804, his explorations resulted in gathering many botanical and other natural history specimens. Although he found his stay at the Derwent River region tedious, he felt that it was worth staying beyond March 1804 when the *Lady Nelson* sailed back to Sydney.

Brown climbed Mt. Wellington twice, collecting certain alpine plants, and noted marine fossils and bivalve molluscs there from the Permian period. Brown's notes about this discovery reveal that he observed the presence of marine fossils at such high elevations. However, this finding did not have the same impact on Brown that it had on Darwin almost 30 years later when Darwin observed marine fossils high in the Andes. At the time, Brown's notes were rather sparse and disorganized because he no longer had an assistant to help with his collecting and note-taking. He had to sacrifice recording more coherent notes to devote his energies to the task of gathering samples. There was little opportunity for Brown to reflect much about the discovery of marine fossils at high elevations.

Later in the century, there was a greater readiness to discuss how species might be capable of change, and additional knowledge about changes in the Earth's crust. Darwin and his contemporaries had the advantage of drawing upon the insights of Charles Lyell and other geologists. Brown may have felt less restrained in expressing his thoughts about the phenomenon he observed on Mt. Wellington. It may have remained in the back of his mind but he remained circumspect on the subject.

Before the *Lady Nelson* sailed in March, it was delayed for a short period, giving Brown a chance to write a letter to Lieutenant-Governor William Paterson (1755–1810) who was based in Sydney. Brown informed him of developments

---

[12] *The Indian and Pacific Correspondence of Banks,* vol. 6, Letter # 193, pp. 337–339. Humphrey, a public servant, sailed as a mineralogist with Collins to set up a colony on the southern coast of Australia. Humphrey searched for minerals and made several journeys along with Brown and with the rest of Collins's expedition to the Derwent River, also looking for freshwater which Port Philip lacked.

since he left Sydney, reporting that the weather—"contrary winds"—had detained him for three weeks. He indicated that he was living on board the *Ocean*. Brown had developed a good working relationship with Paterson because Paterson, in addition to his fine reputation as a soldier, explorer, and administrator, was an adept amateur botanist. With his interests in botany, Paterson provided a sympathetic ear for Brown.

Brown's letter was written from Sullivans Cove on the Derwent River. He reported collecting at most 300 plants and that around 40 were new to him. He indicated some ambivalence about remaining in Tasmania, with one part of him waiting for the opportunity to return "to Port Jackson," but at the same time was motivated to do further collecting, that "on the highest Mountain in the neighbourhood ...[I] observed upwards of 40 plants." He expressed his uneasiness "about his specimens" and asked Paterson to inspect several parcels and help him prevent such "deprivations as mice."[13]

Brown and Humphrey journeyed on the *Ocean* up the Derwent River after Brown posted his letter to Paterson (sometime between March 5th and 7th). Afterwards, Brown revisited Herdsmans Cove where he previously had success in collecting. On February 12, 1804, he recorded descriptions of three plants, in pencil in his diary: *Dodonaea* (Sapindaceae), *Aster* (Compositae), and *Drosera* (Droseraceae).[14] On March 14th, both Brown and Humphrey climbed Mt. Wellington (Table Mountain) again, and Brown noted the geology and the presence of Tasmanian blue gum (*Eucalyptus globulus*).

For the remainder of March and April, the explorations continued in the same fashion. Notable was Brown's discovery of a heretofore-unnamed river—which ran southward from the camp—when he went off on his own, accompanied only by his servant, John William Porter. The river was eventually named after Brown. He discovered it attempting to find the Huon River. Brown felt that he really was not conducting substantive botanical investigations so he had reservations about the usefulness of his work at this point. It seemed to him he was assisting the others in their exploration of the region although he reported to Banks and others at home that he was taking advantage of the opportunity to supplement his previous work.[15]

On May 1st, Brown and his party returned to Sullivans Cove from the Risdon River area. They then journeyed to Prince of Wales Bay on the western banks of the Derwent River, opposite Risdon Cove. Brown took soil samples and observed that those taken from the swampy terrain were "tolerable" but not as good as the soil found in Sullivans Cove. Brown observed the blue gum trees he saw earlier (*Eucalyptus globulus*, an evergreen tree), noting that they were "of moderate size." He also described yellow rice flower, *Pimelea flava*, a shrub that can grow up to six

---

[13] A copy of the letter is published in *Nature's Investigator*, pp. 487–488, and there is a copy in the Natural History Museum, Botany Library, *D.T.C.* 14. 237–240.

[14] *Nature's Investigator*, p. 491. Brown's notes for this period were exceedingly sparse.

[15] Bauer continued collecting in the Sydney area, sometime accompanied by Calley, while Brown was involved in surveying the rugged terrain of Tasmania. *Nature's Investigator* is a reliable source of information for this period. It has skillfully incorporated Brown's fragmentary notes in describing both Brown's experience in Tasmania and Bauer's work in the Sydney area (p. 497).

feet, which was indigenous to Australia. Brown later described this plant in his *Prodromus Florae Novae Hollandiae* (1810). Brown also observed *Asterotrichion discolor* (Tasmanian hemp bush, a dioecious plant Brown incorrectly identified as *Sida didyma*), southern sassafras (*Atherosperma moschatum*), turpentine bush (*Beyeria viscose* which Brown labeled *Terebint baccarat*), and *Dicksonia antarctica* (the Australian or Tasmanian tree fern that Brown identified as *D. australis*).[16]

Problems arose when Brown's servant, Porter, came down with a bad case of "vertigo" and could not proceed further (May 29th–30th). Then Brown fell "among the rocks ... [and] sprained" his foot so badly that he had to stop his climb up Mount Wellington at the end of the month (May 31st, 1804)—Brown continually referred to it as "Table" Mountain in his diary. Because the goal of this journey was to advance Britain's colonial interests and survey the area for mineral deposits, not to particularly study the region's natural history, naturalists like Brown were allowed to "tag along," with the understanding that their interests were subordinate to the overall goals of the project. Therefore, if he was incapacitated for any length of time, there was no one who would be able to help him carry out his work. When he traveled on the *Investigator*, he was able to rely on others if he became incapacitated or was otherwise occupied.

Brown's diary record for May 31st is a more thorough entry but chiefly concerns his geological observations and his ankle injury. His foot was bothering him so much he could not very much travel on his own. Rev. Robert Knopwood (1763–1838), an amateur naturalist and the first clergyman in Tasmania, assisted Brown some-what by shooting birds so he was able to add some ornithological information. Brown remained incapacitated until well into June. As a consequence, there were no diary entries until June 12th.

During this period of time, Bauer remained in the Sydney area, collecting with George Caley—who continued to serve in his usual capacity—in Parramatta, mak-ing illustrations of what he observed. Caley, whom Banks at first was reluctant to engage as a naturalist on the *Investigator*, proved to be a very productive botanical collector. When Banks had reluctantly appointed Caley to the position in 1798 before plans for the expedition on the *Investigator* were firmed up, he was pessimis-tic about Caley's abilities to make any sort of contribution but Caley wound up exceeding Banks's expectations.

After recuperation from his ankle injury, Brown finally left Risdon Cove on a whaleboat, commanded by John Bowen (1780–1827) and was accompanied by Jacob Mountgarrett (1773–1828), a colonial surgeon with whom Brown got on well.[17] They landed on Bruny Island where Brown observed *Asterotrichion discolor* (Tasmanian hemp bush, the dioecious plant he noted the previous month), and then

---

[16] *Brown's Diary*, Natural History Museum, Botany Library. Also in *Nature's Investigator*, p. 503.
[17] Mountgarrett had a dubious reputation however. He was suspected of murder and robbery but Brown got on well with him.

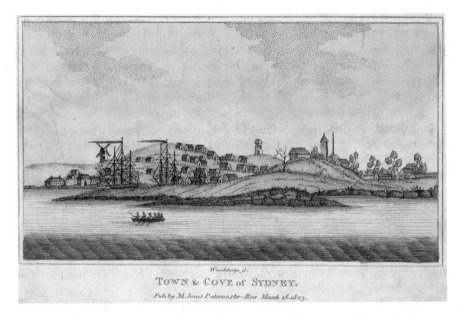

Fig. 9.2 *Town & Cove of Sydney*, pub. by Mr. Jones, Paternoster Row, London, March 18, 1803

moved on to the mouth of the Huon River where he observed mountain pepper, *Tasmannia lanceolata*, called *Wintera didyma* at the time, and *Leucopogon lanceolatus*, an herb commonly known as lance beard-heath. They then explored the Esperance River region. The diary entry for June 17th noted that "the vegetation on" the banks of Port Esperance (Port de l'Esperance) was "uncommonly luxuriant" with blue gum, several other *Eucalyptus* species such as stringy bark, and forest or red mahogany. Brown also reported that someone shot a male platypus.[18]

In the remaining time he had in Tasmania, Brown explored the Derwent River region, visiting Storm Bay, Bruny Island, the mouth of the Huon River, and Risdon Cove. In August, Brown boarded the *Ocean* (on August 8th), and the ship departed Tasmania and sailed the next day (9th) for New South Wales. Brown had compiled an extensive list (handlist) of the plants he observed and collected in Tasmania, he titled "Florula Montis Tabularis."[19] He arrived back in Sydney on August 24th, and when he landed learned that Bauer had left for Norfolk Island several days before. As a result, Brown and Bauer continued their investigations apart from one another. They had been busily conducting their work separately for a total of 15 months, a circumstance not entirely planned. However, their efforts proved to be quite fruitful (Fig. 9.2).

---

[18] Brown previously referred to a female platypus in his letter to Banks a year before (August 6th 1803), *B.L. Add. MS.* 32,439, 104–108.

[19] The complete list is in Brown's diary, with an extremely useful transcription included in *Nature's Investigator*, pp. 520–526.

Work in the Hunter River Area

Banks wrote to Brown—on August 30th—informing him that the French imprisoned Flinders: "Poor Flinders is a Prisoner & I fear not very well treated."[20] Banks
explained the circumstances why this occurred. Flinders did not know that Britain
and France were again at war and after returning to Sydney, he took command of a
29-ton schooner (the *Cumberland*), which would travel back to England. The
Cumberland was not an ideal vessel for such a trip and conditions on board were
uncomfortable; they were crowded and the ship was in need of repair. While crossing the Indian Ocean, Flinders had the ship stop at Isle de France (now Mauritius)
for water and repairs (on December 17, 1803). The French Governor was suspicious
of Flinders, regarding him as a spy, and imprisoned him.

Banks grew apprehensive thus his motivation for writing Brown. Banks suggested that Brown and Bauer should secure passage on the first government ship
returning to England. The treatment that Matthew Flinders received seemed unjust
in view of Banks's beliefs about the universality of science, and by the generosity
he exhibited toward foreign nationals, whether or not the nations were at war,
including the French with whom there was so much animosity. Flinders's imprisonment by the French was matched it seemed by callous indifference on the part of
British authorities. Banks was unable to do much to assist Flinders and his enlight-
ened belief that although countries may be at war with one another, scientists and
nonscientists who were not military officers from different warring countries should
be treated with respect. However, whether it was the fault of communication on his
part or his relative powerlessness, Banks was not able to do anything for Flinders.[21]

Banks informed Brown that the *Calcutta* had arrived with "twelve kegs" of specimens. He informed him that the seeds Brown had sent to Kew "produced some
curious plants." Banks had just learned from John Allen that Peter Good died, illustrating the time lag and slow communication in the world of the early part of the
nineteenth century: "No mention is made in any letter that I have heard of poor Peter

---

[20] Letter from Banks to Brown, August 30, 1804, original in British Library, *B.L. Add. MS.* 32,439.
95; copy in Natural History Museum, Botany Library, *D.T.C.* 15: 84–86. Also copy included in *The
Indian and Pacific Correspondence of Banks,* vol. 6, Letter # 210, pp. 373–374.

[21] Elise S. Lipkowitz examined the cosmopolitan ideals in Europe at the time of the French
Revolution. She saluted Banks's role in assisting the French naturalist, Jacques-Julien Houtou de
Labillardière (1755–1834), to retrieve his natural history collections in the 1790s, which were
seized by British forces as war booty. Banks believed that "sciences were never at war," but these
high-minded ideals were put to the test later on, and were not helpful in freeing Flinders. Lipkowitz
indicated that the French and British approaches reveal a "new scientific nationalism" that ran
counter to the earlier attitudes exhibited by both nations, which were driven by French scientific
universalism and British liberal scientific improvement. Lipkowitz, "Seized natural-history collections and the redefinition of scientific cosmopolitanism in the era of the French Revolution," *The
British Journal of the History of Science* (2014) 47 (1): 15–41.

Good, we are told by M^r· Allen the Miner who returned in an Indiaman, that he dies at Port Jackson, which I fear has been the case."[22]

Brown spent his remaining time in Australia studying the flora and fauna in the Sydney area, taking advantage of the opportunities this island continent afforded him. Before going to Hunter River in October, he had a very productive visit with George Caley in Parramatta in September. Brown valued Caley's insight and experience with the native Australian plants.[23] Brown walked from Sydney to Parramatta— today a suburb of Sydney, about 14 miles west of the Sydney business district—on Friday, September 1804. His diary listed many different plants he came in contact with, including red bloodwood, Sydney red gum, different species of ironbarks, forest and swamp mahogany, turpentine, wire weed, as well as several species of bats (Chiroptera) who were not "blood-feeding," and birds from the Phasiandidae— including lowland pheasants—and Turnicidae—a family of birds made up of species of quail, particularly the button quail.

Brown spent several days discussing and sharing his botanical observations with Caley, and he was particularly impressed with the flora Caley collected from the Hunter River region of New South Wales. Brown was determined to travel there himself and conduct his own observations and collect in this area. During the period of October–November 1804, he spent considerable time exploring this region. He sailed on the schooner *Resource*—which was carrying provisions for the settlement there—for King's Town (Kingston).

On the morning of October 9th, Brown put a few boxes containing paper (paper that could hold the seeds he collected) on the ship that was to depart for Coal River (an alternate name for Hunter River). He had to sleep a night or two on board the *Resource* because the weather was not suitable for departure (the wind was quite fierce). Finally, on October 11th the schooner departed from Sydney and arrived in Newcastle the following day (October 12th). He recorded that initially they were nearly "abreast" Hunter River (he sometimes referred to it as Hunters River). Owing to the ebb tide and high seas, they were unable to dock until the evening.[24] When he went ashore the next day, he observed a few plants he had not previously seen such as *Baeckea ramosissima* (a shrub in the Myrtle family) and *Centaurea* (a thistle-bearing plant in the Asteraceae that Caley observed in the mountains but also found along the shore). He visited the coal mine there and walked around, noting the sandy

---

[22] Letter from Banks to Brown. Because of the slowness of the mail, Brown and Bauer did not see this letter. By the time it arrived in Australia almost a year later, on June 7, 1805—carried by the *Argo*—Brown and Bauer were on their way home, *British Library Add. M.S.* 32439. 95; copies in *D.T.C.* 15: 84–86 and *The Indian and Pacific Correspondence of Banks,* vol. 6, Letter # 210, pp. 373–374.

[23] *Nature's Investigator* indicates that "it is clear Banks' collector [Caley] was now taken seriously," noting that Brown rarely praised anyone but his notes "are a notable compliment to Caley," p. 528. Also see Joan B. Webb *George Caley, Nineteenth Century Naturalist: A Biography* (Chipping Norton, New South Wales: Surrey Beatty & Sons, 1995).

[24] *Nature's Investigator*, p. 542.

condition of the soil and that the hills were covered with shortgrass but the area was devoid of trees and shrubs.

Brown spent a few days searching the Newcastle area, again finding little in the way of shrubs or trees. On October 16th he made his first trip up the Hunter River. He explored the banks of the river, in the area where there was extensive cutting of cedar, and observed gray mangrove (*Avicennia marina* var. *australasica*) there as well.[25] He found that the banks were quite swampy and covered with trees, particularly a "remarkable" species of *Ficus* (*Ficus macrophylla*). This tree is commonly known today as the Moreton Bay fig. It is native to the east coast of Australia and is an evergreen banyan tree belonging to the Moraceae that has imposing buttress roots. He also observed swamp oak, an evergreen tree common in the coastal areas of Australia (*Casuarina glauca*), prickly leaved paperbark, and various birds such as black swans (*Cygnus atratus*), bell miners (*Manorina melanophrys*, also called bellbirds), and dollarbirds (*Eurystomus orientalis*).

Brown found this experience invaluable in that he had the opportunity to observe organisms that were unique to the Australian east coast. He spent the week exploring the Hunter River and a tributary, the Patterson River, returning when provisions became scarce. On October 31st, he reported difficulties with natives who he thought were friendly initially. However, the natives lingered about and thus Brown ordered the provisions be packed on the boat with the natives trailing them. They became hostile when they saw that the boxes and other bundles were out of their reach, and they became irritated and one struck Brown's servant because Brown and his party were unable to affect an easy escape due to the low tide of the river at that point. After Brown and his men fired some shots, the natives dispersed, and they did not see them again.[26]

The Return Home

After spending most of November in the Hunter River area, Brown returned to Sydney, arriving there on November 20, 1804. He spent the Australian summer (December 1804–January 1805) in Sydney and the nearby communities. Before leaving for a trip to the Grose River area, he wrote to both Banks and Greville (on December 12th) to inform them of his progress since he last wrote. He informed Banks that he had shipped 12 puncheons (large wooden casks) of plants and 4 boxes of seeds on the *Calcutta*, but the animals and minerals were not yet sent because he was not at Sydney to pack them. Brown confided that he was not in good health:

---

[25] Brown's diary often erred in giving the plants' scientific nomenclature and the names of the geographical places such as rivers. Vallance's work is very helpful in sorting out these errors, and providing the correct names.

[26] *Nature's Investigator*, pp. 553–554. A checklist of the plants Brown collected are in *Plantae Novea vel Noviciae Florae N Hollandiae, in vicinitate Newcastle & ad ripas flaviorum Hunter's Paterson's & William Rivers observatae, Oct–Nov 1804*. Also in *Nature's Investigator*, pp. 557–560.

"For some months past I have not enjoy'd good health. I have often been so weak as to be incapable of undertaking any laborious excursion." He did report that he collected for some time in Van Diemen's Land, after having left Port Jackson at the end of November, 1803 (28th), on the "Colonial tender, Lady Nelson, hoping to be able to add largely to my collection in Van Dieman's Land, but not expecting to be absent more than eight or ten weeks. From various unavoidable circumstances, however, my stay was protracted to nine months, of which time a very great part was entirely lost …. Van Diemen's Land is by no means so rich in plants as I expected to have found it …." He also reported that Bauer did not accompany him to Van Diemen's Land but since returning from Van Diemen's Land he visited "Hunter's River & examind all the branches" and added about 50 species of plants to his list. He also wrote that he was anxious to hear from Banks about Flinders "who left this in so small a vessel that we are not without fears for his safety."[27] Obviously, Brown had not yet received Banks's letter of August 30th that contained the news about Flinders's capture.

Brown's letter to Greville—the first Earl of Warwick, a patron of science as well as a mineralogist and a horticulturist—contained substantially the same information expressed in his letter to Banks. Noting Greville's interest in mineralogy, Brown informed him that he had become "acquainted" with Adolarius William Henry Humphrey. Humphrey was Greville's disciple, and had gained his position as mineralogist in New South Wales with Greville's assistance. He informed Greville that his visit to Van Diemen's Land fell far short of his expectations, even though he spent considerable time there.[28] Actually, Brown was somewhat harsh in his self-assessment. His negative comments obscure the fact that he had made the most of the opportunity of exploring since he left the *Investigator* (particularly his work in Tasmania).

Brown visited Caley at Parramatta again to see his "Blue Mountains" collection (on December 16–17, 1804). Caley had collected specimens during his trip to the Carmarthen Hills (today known as the Blue Mountains) in November. Caley reported that this trip took 20 days (from the 3rd to the 23rd of November), describing it as "the most laborious man ever went to."[29] Caley wrote a lengthy account of this trip for Banks, and entrusted Brown to carry it back to Banks when Brown finally left for England in 1805.[30] Because there was a great deal of material to examine at Caley's home, Brown's observations took several days. After studying Caley's collection, Brown set out to explore different places in the region, the ponds (Burralow Creek) near Hawkesbury Road where he observed *Fusanus crassifolia* (New South Wales sandalwood) and *Damasonium minus* (star fruit or rice weed, found in ponds in NSW), Badgery's Farm, where he spent Christmas, and then

---

[27] Letter from Brown to Banks, December 12, 1804, *D.T.C.* 15: 185–188. Also in *The Indian and Pacific Correspondence of Banks,* vol. 6, Letter # 220, pp. 387–390.

[28] Brown's letter to Greville, December 12, 1804, *British Library, Add. MS* 32439 ff. 159–160.

[29] Joan Betty Webb, *George Caley, Nineteenth Century Naturalist, A biography* (Chipping Norton: S. Beatty & Sons, 1995), p. 69.

[30] The manuscript is now preserved in the Natural History Museum in London.

explored the Grose River. He spent the remainder of the time making a thorough study of the area and compiled a Grose River plant list (from December 1804 to January 1805).[31]

From Sydney, he wrote to Banks a detailed letter on February 21, 1805, that his health had not gotten much better but had continued to collect, explaining that many specimens were lost when the *Porpoise* was wrecked but had sent a "small box … [which] contains the few seeds collected chiefly in that excursion." He enumerated in some detail some curious discoveries such as a "genus of the Malvaceae near Pavonia but having only 5 stamina." He also reported finding "the finest species of Mimosa I have hitherto seen" in New Holland and that he had sent home on the *Lady Barlow*. He again informed Banks that Bauer had not yet returned from Norfolk Island. As a consequence, Brown had not seen him for some time, and there was no word about the surgeon-naturalist George Bass (1771–1803), who later was presumed to be lost at sea.

In a postscript to the letter, Brown provided a list of a "General Account of a Collections of Natural History of New Holland formed during the Voyage of the Investigator & subsequently at the settlements in New South Wales & Van Diemen's Island." The list included "Specimens of Plants of the South Coast/from King George III^{dh} S^d to Bass Strait/Seven Hundred Species, Of the East Coast from Sandy Cape to Cumberland Isles, Five Hundred Species, Of the North Coast from Endeavour's Strait to Arnheim's Land, Five Hundred Species, Of the vicinity of Port Jackson, One Thousand Species, Of Van Dieman's Island, Seven Hundred Species, and Of the Island of Timor, Two Hundred Species."[32] The concise summary of the work he had accomplished was followed by Brown's estimate of the expenses he believed would be incurred in transferring these collections from Liverpool to the Custom House in London when he finally arrived home.

In the early part of 1805, Brown began to make preparations to return to England. He was reunited with Bauer on March 11, 1805, when the repaired *Investigator*— which was now employed for explorations in the coastal areas near Sydney—arrived from Norfolk Island. They had not seen one another since November 1803 when Brown left for Bass Strait and Van Diemen's Land on the *Lady Nelson*. Brown traveled again to Parramatta to see Caley. Bauer may have accompanied him; his completed drawing of *Chiloglottis reflexa* (Autumn Bird Orchid) is marked Sydney-Parramatta, March 1805, so it appears that he went with Brown on this visit to Caley. Plans were now being made by Governor King to have Brown and Bauer return to England because the growing tensions between the European powers had extended to even the remotest outposts. Because the officers on the brig *Harrington* had assumed that hostilities resumed between Spain and Britain, the *Harrington* seized two Spanish ships. The Spanish and their allies considered this an act of piracy, so it was deemed prudent for the naturalists to return as soon as possible.

---

[31] *Nature's Investigator*, p. 573.

[32] Letter from Brown to Banks, February 21, 1805, *Brown Correspondence* 3.123. Also a copy in *The Indian and Pacific Correspondence of Banks,* vol. 6, Letter # 240, pp. 427–430.

Governor King decided that the newly refitted *Investigator* should return to England (on March 18th). The upper deck of this ship had been removed and the original lower deck became the only deck with half of its original tonnage. It was re-rigged as a brig with two masts rather than its original three. With these changes, King believed that the ship could accommodate Brown and Bauer and their specimens. In the meantime, Brown was on his way to the Hawkesbury River area because he felt it necessary to revisit the area he explored previously with Good in June 1802. King determined it would be best that Brown and Bauer should return with their collections to England on the *Investigator* but Brown was not happy with the prospect of again putting his faith in the vessel that previously proved to be unseaworthy. On March 23rd, Brown wrote his objections in a letter to King, emphasizing that the dampness on board would compromise his specimens.[33]

King went ahead with his plans nonetheless, while Brown visited Caley for the last time. Caley prepared the plants for Brown for transporting them to England aboard the *Investigator* although he was not entirely confident that the ship would be able to carry these plants safely back to Britain. King, in his letter to Banks (May 20, 1805), indicated that Caley had nothing or chose not to send anything on the *Investigator*. Caley's almost feverish preparation of the material contradicts this assertion. King made this claim because Brown left the garden that had been rescued from the *Porpoise* in Caley's care.[34] While this was happening, the so-called *Harrington* affair continued to have an effect on the planned departure of Brown and Bauer. The Spanish vessel, *Estramina*, arrived in Sydney and at the same time word about Flinders's situation revealed that he was being held prisoner with no hope of rescue from his plight (Fig. 9.3).

King wrote to Brown at the beginning of May (May 9th), a few days after Brown conducted his final collecting trip in the Sydney area (May 5th). A day earlier (May 8th), Brown inspected the *Investigator* and was not particularly comfortable with plans to transport his specimens in the storage of the *Investigator*. King indicated to Brown that he was "averse to interpose any authority in the direction of your collection, otherwise than in facilitating its safety, it will remain with you and Mr. Bauer to decide on the propriety of yourselves and collection going by the Investigator." He added that it was not in his power to determine when another opportunity within his authority presented itself.[35] Brown promptly answered King (on May 11th), telling the governor that he observed no improvement in the conditions, "my opinion of its absolute unfitness for the reception of specimens remains unaltered."[36] Brown wanted some assurance that his material would not be just stored in the ship's hold

---

[33] Letter from Brown to Gov. Philip Gidley King, March 23, 1805, *Historical Records of the New South Wales* (*HRNSW*) 5: 580; also in *Nature's Investigator*, p. 583.

[34] Letter from King to Banks, May 20th, 1805, *D.T.C.* 16. 26–30). Also a copy in *The Indian and Pacific Correspondence of Banks,* vol. 7, Letter # 23, pp. 41–44.

[35] Letter from King to Brown, May 9th, 1805. *HRNSW* 5: 616–617; also reprinted in *Nature's Investigator*, p. 587.

[36] Letter from Brown to King (on the condition of the Investigator), May 11th, 1805, *HRNSW* 5: 619; also reprinted in *Nature's Investigator*, pp. 589–590.

Fig. 9.3 Governor Philip Gidley King, by an unknown artist, State Library of New South Wales

but he would be given a small cabin as King promised (without consulting the man who would Captain the ship, Captain Kent on the matter). Governor King wrote to Banks on May 20th, informing him that the *Investigator* had been repaired and Brown and Bauer and their collections should return on the ship. However, Brown's collections that were rescued from the Porpoise would be left with Caley.[37]

In spite of his misgivings about conditions aboard the *Investigator*, Brown realized that he would have to settle on the best possible terms he could negotiate because he determined from King's firm stance that there would be no other possibility of returning home. If Brown and Bauer rejected it, King made it clear that he would not exert himself further in assisting Brown gain passage home. King wrote to Brown on May (16th) assuring him that the accommodations with regard to the cabins and his collections should put him "at ease."[38] With no alternative and a great deal of trepidation, Brown reluctantly agreed to go on the *Investigator* under the command of Captain Kent. He stowed his herbarium and the least perishable part of his collection in the ship's hold. The living plant material was left at Parramatta under Caley's care (Fig. 9.4).

---

[37] Letter from King to Banks, May 20th, 1805. *D.T.C.* 16: 26–30.

[38] Letter to King to Brown, May 16th, 1805, *HRNSW* 5: 626; reprinted in *Nature's Investigator*, p. 591 and in *The Indian and Pacific Correspondence of Banks,* vol. 7, Letter # 23, pp. 41–44.

Fig. 9.4 Robert Brown, likely drawn after his return from his exploration in Australia, 1837, line engraving by C. Fox after Henry William Pickersgill, Wellcome Collection Library no. 1426i

The *Investigator* departed from Australia on May 23rd. After more than four difficult months at sea, it arrived in Liverpool, England, on October 13, 1805, because the *Investigator* was in no shape to negotiate the tricky conditions in the English Channel. There was little acclaim when Brown arrived home. Brown obviously was glad to be back in England after being away for more than four years (July 1801–October 1805) and a difficult passage on a ship that was barely seaworthy. He did not receive the outpouring of notices in the newspapers and other contemporary periodicals in same fashion that Park received when he returned from his first journey to Africa.[39] Brown may have felt disappointment about his rather unceremonious arrival home, but he had much to consider. He had to take care of the large amount of specimens he had brought back, and determine how to best catalogue them along with the seeds and other specimens that he had already sent to Banks. Most of all, he had to think about how he would spend the next part of his life and career.

---

[39] *Nature's Investigator* reprints a letter giving an account of Brown's arrival. It was written by someone identified only as P.M.T., who wrote a letter to the *Atheneum* no. 1603 July 17, 1858, stating that the ship and Brown and Bauer's arrival were "afforded" a "ready welcome by sympathetic friends" although it claims that this was "a fact Brown left unnoticed," p. 599. This remembrance ran counter to the information contained in the letters exchanged between Brown and King at the time. The tone of the letter indicated that it was written to defend or absolve King in the matter, and support King's contention that Brown was responsible for his traveling back with his specimens on the *Investigator*.

Chapter 10
Prodromus, Florae Novae Hollandiae

Contents

When Brown and Bauer arrived in England in October 1805 on the repaired *Investigator*, they had nearly 4000 species of flora in its hold (3200 from Australia and 200 from Timor), a good sample of Australian fauna, and Bauer's excellent illustrations as well.[1] When they landed in Liverpool, Brown wrote to Banks (on October 13, 1805), summarizing his work and journey home. Describing his passage home as "tedious and uncomfortable," he explained that it was impossible to take the entire garden. He reported that Governor King assured him that it would be conveyed on a subsequent voyage. He enclosed a list of packing cases containing the material he collected, and listed the contents of each case. He even went to the trouble of listing the chests of drawers containing his "private luggage."[2]

In retrospect, Brown's complaints about his voyage were not excessive or unwarranted. He exhibited remarkable fortitude in view of the effects that travel and exposure to the harsh climate, particularly heat, had on him. Additionally, he had nagging worries about how the specimens of flora and fauna would be handled. His health suffered as a result of more than five years of the often harsh conditions he endured, and spent the final trip home confined to the cabin of a relatively small vessel not

---

[1] 236 drawings of Australian plants drawn by Bauer are in the British Museum (Natural History). See William T. Stearn, *The Australian Flower Paintings of Ferdinand Bauer*, Introduction by Wilfred Blunt (London: The Baslick Press, 1976). Also see "Ferdinand Bauer's Drawings of Australian Plants, *The Journal of Botany* (1909) 47:140–146.

[2] Letter from Brown to Banks, October 13, 1805, *British Library*, Add. *MS*. 32439 ff. 183–184. However, there were problems and Caley retained possession of the garden. There is no evidence that the garden ever was transported to England. *Nature's Investigator* indicates that Brown did not seem overly concerned with this and seemed more concerned with the fate of his herbarium, pp. 599–600. A copy of the letter is in *The Indian and Pacific Correspondence of Banks,* vol. 7, Letter # 43, pp. 77–79.

© The Author(s), under exclusive license to Springer Nature Switzerland AG 2021   141
J. Schwartz, *Robert Brown and Mungo Park*, Memoirs of The New York
Botanical Garden 122, https://doi.org/10.1007/978-3-030-74859-3_10

really suited for such a journey. He bore this hardship with his usual serenity, and was most eager to study the vast material he had accumulated. He planned to identify the plant specimens, examine their anatomy as thoroughly as possible, and determine their position in the plant kingdom. In doing so, sometimes he found it useful to rename them to be consistent with the information he gathered on the organisms he observed and collected.

Brown indicated to Banks that it was solely Governor King who urged Brown and Bauer to travel back to England on the *Investigator* along with their collections: "Governor King recommended her [The *Investigator*] as the best conveyance for the collection that was likely to offer." Brown expressed his dissatisfaction with the conditions on the *Investigator*, indicating however that during the voyage, "the plants were carefully examin'd" and kept dry as could be on a vessel that was in such a "wet state." He urged that the specimens should be removed on shore "earnestly" begging "that they may not again be put on board the Investigator for the purpose of being brought round to Portsmouth." To avoid this, he suggested that the collection be removed as soon as possible from the ship and be transported to London over land.

Captain Kent, who commanded the voyage home on the *Investigator*, traveled ahead to London to directly report to the Admiralty. There, he asked that the cases with Brown's collections to be unloaded as quickly as possible. Following the instructions Kent received from the Admiralty, they were unloaded (on October 29th). However, because the "custom house" in Liverpool took special interest in the 36 cases, Brown's departure from Liverpool was delayed because customs in Liverpool would not release them.[3] They sat in Liverpool much to Brown's dismay. Banks was unaware that Brown was having difficulties with customs in Liverpool because he was at his estate in Lincolnshire (Revesby Abbey). In view of the fact that Brown was an employee of the Admiralty, it may have been better if he had directed his request to them initially.

On October 19th, Banks had written to William Marsden (1754–1836), Secretary of the Admiralty, from his estate in Lincolnshire, and vouched for Brown's trustworthiness and belief that the specimens and sketches should be properly treated.[4] This letter merely facilitated the removal of the cases from the ship and did little to ease the situation regarding customs in Liverpool. Brown was growing increasingly frustrated in Liverpool, and finally wrote to Jonas Dryander, Banks's librarian, on October 29th, asking for assistance. He indicated his fear that he was likely to remain in Liverpool for six months or longer. This may have been an overstatement but it reflected Brown's irritation and frustration. He informed Dryander that "I

---

[3] There is no plausible explanation for this delay except perhaps bureaucratic foot-dragging. Brown's letter to Banks, October 13, 1805, *British Library, Add. MS.*32439 ff. 183–184 & *The Indian and Pacific Correspondence of Banks*, vol. 7, Letter # 43, pp. 77–79. Captain Kent's letter to William Marsden, Secretary of the Admiralty, October 13 or 14, 1805, published in *Nature's Investigator*, pp. 598–599.

[4] October 19th letter from Banks to Marsden, *D.T.C.* 16:149–150 & *The Indian and Pacific Correspondence of Banks*, vol. 7, Letter # 44, pp. 79–80.

requested Cap Kent if he found that Sir Joseph Banks was in the country, [meaning his estate in Lincolnshire] to state at the Admiralty the necessity of removing the collection from Investigator, and forwarding it to London by land." Brown clearly believed that Banks had the ultimate authority in this matter. He acknowledged that although removal of his specimens had been accomplished, Kent failed "to make a regular application for a dispensing order, without which it cannot be removed from the Custom house."[5]

Apparently, Brown and Banks—perhaps with Brown's continual prodding—finally were successful in getting the materials through customs. Now they had to be transported by coach (wagon) to London and Brown needed to inform the Admiralty what had transpired. In a memorandum Brown sent to Secretary Marsden on November 5th, he stated that it would be better for the crates to be transported to London by wagon, rather than by taking them by the waterways of England, a less expensive mode of transport.[6] Brown found it necessary to write a memorandum to Marsden because when both he and Bauer arrived in London on November 5th, Marsden did not take the trouble to see them. Brown explained to Marsden that "Robert Brown, Botanist & Ferdinand Bauer Painter of Natural History, belonging to His Majestys Ship Investigator, have the honor of acquainting M[r] Marsden, for the information of their Lordships, of their arrival in London, and that the Collection of Natural History brought to England in the Investigator, was sent by Waggon [sic.] from the Custom house of Liverpool on Friday the first inst, addressed to the Kings Ware House, Custom house, London. They beg leave to state their great anxiety that the collection may not be detained or opened at the [London] Custom house, fearing that from the very wet state the Investigator was in during the greater part of her passage from New South Wales, it may have already sufferd considerably & that therefore it will be much endangered by being carelessly handled."[7]

The reception and great acclaim Park had received upon his return from his first trip to Africa are in stark contrast with what Brown and Bauer had to endure, i.e., the weeks they spent in Liverpool, waiting to get clearance for the valuable material they collected, and then suffering indifference when they arrived at the Admiralty. Also, Banks did not make an effort to shorten his stay in the Lincolnshire country-side and greet the naturalists. Moreover, Marsden did not immediately reply to Brown. This indifference speaks loudly about the priorities of the Admiralty and what they deemed important. In spite of this, Banks did urge the Admiralty to extend Brown and Bauer's employment, and they remained on its payroll, thereby affording them some financial security in the meantime.

Dryander wrote to Banks and gave an account of the treatment Brown and Bauer received at the Admiralty. He reported, "M[r] Brown and M[r] Bauer made their

[5] Letter from Brown to Marsden, October 29, 1805, *B.L. Add. MS* 32439 f. 186 r, v.

[6] *B.L. Add. MS* 32439 f. 188 r, v.

[7] Memo from Brown to William Marsden, Secretary of the Admiralty, on November 5, 1805, in Brown's own hand, *British Library, Add. MS* 32439 f. 188 r & v. It is clear that Brown wanted to get down on record the rather shoddy treatment he and his collections had received, and, once again, his displeasure over the conditions on the *Investigator.*

appearance here," explaining because Captain Kent had not given any direction to Brown, he "therefore sent the collections by wagon, as the most safe conveyance .... Mr Brown went to the Admiralty ... but they could not see Mr Marsden, & as they waited there for 2 or 3 hours."[8] Dryander wrote to Banks because he felt that Banks would not return from his country retreat to take care of the situation himself. In spite of the cavalier treatment Brown had received, the Admiralty was not disinterested with the work Brown had undertaken. They just did not consider Brown's work of paramount importance and were still aglow from the great naval victory the British had achieved at the Battle of Trafalgar (on October 21, 1805).

However, Brown remained on the payroll of the Admiralty until 1810, and was well compensated for his efforts because Banks had suggested to officials that Brown should publish a full account of the material he collected and be properly paid for his efforts. The material Brown brought back with him framed the work he accomplished for the rest of his career, eventually spurring breakthroughs in cytology, palynology (study of pollen grains), taxonomy, and biogeography.[9]

Brown's work on the Rhamnaceae is an example of work he accomplished soon after his return. Rhamnaceae are a group of flowering plants belonging to the buckthorn family, a medium-sized plant family consisting of mainly trees and shrubs and some vines with 900–1000 species. Plants belonging to the Rhamnaceae were notable for their woodiness. It was also found that plants in this group had their stamens (male part of the flower) alternate with their sepals. Prior to Brown's work in Australia, there were no known Australian species belonging to this family. Brown collected 31 species of Rhamnaceae, most of them from the southern and eastern coasts of Australia. These previously had not been described, and Brown added plants from Rhamnaceae he collected from the Sydney area and Tasmania, thereby adding to the abundance of specimens he gathered belonging to this plant taxon.[10]

---

[8] Letter from Jonas Dryander to Joseph Banks, November 7, 1805, Banks Correspondence f. 150. Also there is a full transcript of the letter in *Nature's Investigator*, p. 605, and in *The Indian and Pacific Correspondence of Banks,* vol. 7, Letter # 48, pp. 104–105. In fairness to the Admiralty, they had other things on their mind. Britain had won a great victory at the Battle of Trafalgar on October 21, 1805, so the details concerning Brown's investigations commanded less of their attention at the time.

[9] Phyllis I. Edwards indicates that Banks wrote to Marsden on January 3, 1806, and presented Brown's rough estimate of the plants he collected and where he collected them, the animal species collected, and minerals that he brought back, "Robert Brown (1773–1858) and the natural history of Matthew Flinders' voyage in H.M.S. Investigator, 1801–1805," *Journal of the Society for the Bibliography of Natural History* 7 (1976): 385–407, 403–404.

[10] Brown, "General Remarks, Geographical and Systematical on the Botany of Terra Australis" in *The Miscellaneous Botanical Works of Robert Brown*, ed. by John J. Bennett. (London: Ray Society by R. Hardwicke, 1866–68), vol 1, pp. 1–89, p. 26. [Reprinted from Matthew Flinders, the appendix of *A Voyage to Terra Australis*. In two volumes and an atlas in the third. (London: G. and W. Nicol, 1814).] Jürgen Kellermann discusses the taxonomic history of Rhamnaceae in the first half of the nineteenth century and Robert Brown's contributions to the subject, "Robert Brown's Contributions to Rhamnaceae Systematics," *Telopea. A Journal of Plant Systematics* (2004) 10 (2): 515–24. The issue was devoted to the symposium, "Robert Brown 200," held in the Royal Botanic Gardens of Sydney, Australia, 2004.

As a result of his labors and his success in bringing so many specimens back with him, Brown was elected clerk, librarian, and housekeeper to the Linnean Society of London with a salary of £100, beginning at the end of the year (December 17, 1805). The duties involved with this position were not too onerous, enabling him to work on the materials for exhibit on behalf of the Admiralty, efforts that allowed him to earn a salary of approximately £300–400 per annum. This income permitted Brown the freedom to focus work on the large number of specimens and drawings that were the result of his explorations without having the burden of figuring out how to best support himself.

Brown now had the opportunity to pull together the materials that were the fruits of his journey. His first-hand experience observing many previously undescribed species of plants—particularly from the Rhamnaceae as well as the Proteaceae, another group well represented in Australia—gave him impetus to move from conducting exclusively observational work to challenging the accepted views in British plant taxonomy. In addition, he was able to focus on the relationships among the different species of flora he collected. Most importantly, Brown now had gained some measure of financial security from his position at the Linnean Society and at the Admiralty.

With a reliable source of income and no direct need to depend on the military—he remained on the payroll of the Admiralty—or eke out a living as a country doctor to support himself and his scientific work, Brown felt relatively secure. Brown therefore achieved the status of a professional scientist, one of the earliest scientific professionals, forging a path similar to the one later followed by Huxley, who also went on to make landmark changes in taxonomy, in the latter's case, chiefly in invertebrate zoology.

Brown summed up his journey in a January 12, 1806, letter to his friend and colleague, J. E. Smith, President (and founder) of the Linnean Society. He informed Smith that he had been elected Librarian, a post "more suited to my inclinations than abilities" and modestly explained that "our expedition as you already know has not been uniformly prosperous .... All my best specimens of the South Coast as well as a garden nearly filled with new interesting species ... perished in the Porpoise & a second garden. I have been obliged to leave behind at Port Jackson .... Under such disadvantages we think ourselves tolerably fortunate in being able to bring home about 3000 species .... Mr. Bauer whose abilities ... you are well acquainted with has made about 1600 drawings."[11]

---

[11] Letter from Brown to Smith, January 12, 1806, *Linnean Society* 2:145. Smith had investigated the genus *Cryptandra*, a shrub in the Rhamnaceae family, several years before Brown's departure to Australia. He delayed presenting a full description of *Cryptandra ericoides*—found in heaths in Australia—until 1808, assisted by the annotated list of Rhamnaceae compiled by Brown. Ferdinand Bauer was extremely prolific in the amount of drawings he produced. During their voyage together, Brown paid tribute to Bauer, noting his accuracy and productivity. Several decades ago (1997–1998), there was an extensive exhibit of Bauer's drawings that he made in Australia at the Focus Gallery of the Museum of Sydney, highlighting his proficiency, found in *Ferdinand Bauer (1760–1826), An Exquisite Eye: The Australian Flora & Fauna Drawings 1801–1820 of Ferdinand Bauer*. Peter Watts, J. Anne Pomfrett, David Mabberley (Glege, N.S.W.: Historic Houses Trust of New South Wales, 1997).

In spite of his usual misgivings and wariness, Brown was being unduly modest about the value of the work he conducted during his more than five-year journey. He soon demonstrated the impact his experience had on his development as a naturalist. Brown began to think about the relationships of the great variety of organisms he encountered. He had full access to the herbaria of Banks and Smith (at the Linnean Society). In addition to his collections from Australia and the nearby regions, he had specimens from South Africa, including many species from the Proteaceae. This family of plants, largely indigenous to the Southern Hemisphere, was a group of tropical and subtropical evergreen shrubs and trees including *Macadamia* and *Banksia*. Brown discovered that there was a total of roughly 75 genera and 1000 species of plants belonging to the Proteaceae. He noted similarities between the specimens of Proteaceae from the South African region and those from Australia and its surrounding area.

He continued to inform Smith about his progress. One such letter, written several years after, informed him about his success. Although they did not often see one another, Brown's relationship with Smith was not just a professional one. He regarded Smith both as a friend and a mentor. While informing Smith about his work, Brown continued to express his concern about Smith's wife because her health had not been good. This was the reason why Smith was unable to be at the Linnean Society regularly so Brown's letters served as a means of communicating with his friend and associate. Brown wrote that he was "concerned to learn that Mrs. Smith's state of health does not permit you to come to London as soon as you intended & sincerely hope it may not continue to prevent you being at the Linnean Society's 2nd meeting of April." He reminded Smith, "At present the Society depends entirely upon you for papers." Because of that, Brown indicated that there were not enough "materials for the 9th vol. but if there was any room, Brown had "some intention of offering an essay on Holoraceae .... " (a family of plants including beets and rhubarb). He indicated to Smith that although it was a family that was "far from unimportant in an economical point of view, [it] has been generally neglected by Botanists." Brown indicated that he wanted to examine "the greater part of it endeavouring to determine my new Holland plants belonging to it." (Fig. 10.1)[12]

George Hibbert (1757–1837), a merchant and slave ship owner as well as an amateur botanist, sent his collector, James Niven, to South Africa to bring back rich samples of the Proteaceae and other plant groups. Among the specimens Niven collected from the Cape region of South Africa were seeds of *Nivenia corymbosa* (blue stars) that Hibbert intended to germinate in England. Hibbert's gardener, Joseph Knight, was one of the pioneers in germinating seeds belonging to members of the Proteaceae. Brown named the genus, *Nivenia*, in honor of the collector's fine work. Eventually, *Nivenia corymbosa* was placed in the Iridaceae.

The plants collected by Niven were immeasurably helpful to Brown when he sorted out the specimens he personally collected in South Africa and Australia. The live material Brown brought back included seeds from 16 plants that previously had not been germinated in England. While examining this material, Brown realized

---

[12] Letter from Brown to Smith, April 1, 1808, *Linnean Society* 2:147.

Fig. 10.1 Sir James Edward Smith, with a vignette of the pursuit of the ship containing the Linnean collection by order of the King of Sweden, 1800, by W. Ridley after R.A. Russell, Wellcome Collection Library

that he had further justification to re-evaluate the Linnean system of taxonomy, a system that continued to dominate plant systematics in Britain but not universally accepted in other parts of Europe, notably France. Because of the artificiality of the Linnean system—i.e., artificial classification relies on one or a few characteristics (in the case of the Linnean system, the reproductive parts of the plant)—Brown was already drawn to the more natural system of de Jussieu (embraced by French naturalists), which considered a larger range of characteristics. The vast amount of material Brown assembled gave him sufficient justification in reassessing the Linnean system, and enabled him to better grasp its inadequacies.[13]

Additionally, Brown made some impact on zoological studies. The male wombat that he brought back was given to William Clift, a Conservator of the Hunterian Museum at the College of Surgeons. Clift turned over the creature to the care of Sir Everard Home (1756–1832), a surgeon and amateur naturalist, who published work on the anatomy of humans and other animals. Home kept the animal for 2 years, observing its habits and behavior. When it finally died, Home dissected it and an engraving of its stomach was published in the *Philosophical Transactions* in 1808.[14]

Brown collected a total of 23 mammals, including five kangaroo species, a platypus, and an echidna (a spiny anteater), the latter two belonging to the monotreme order of egg-laying mammals. He also collected three bats, 217 birds, 33 reptiles and amphibians, 39 fish, and 29 invertebrates, not including insects. Brown never worked on classifying the birds and other vertebrates he brought back, organisms that he had studied and sometimes sketched. The same was true for the invertebrates he had observed and collected. He was primarily focused on plant life, and he regarded the other forms of life of secondary interest.

In any event, the vertebrates unfortunately did not survive, meeting the same fate as the wombat he brought back although in that particular case, the creature lasted 2 years. Bauer's drawings of these organisms provided a good record, particularly of *Acanthaluteres brownii* (Brown's leatherjacket, a spiny-tailed file fish native to the Indian and South Pacific oceans and later described by Sir John Richardson, a well-known Arctic explorer as well as a mentor to Thomas Huxley). *Acanthophis brownii*, a member of a group of extremely poisonous elapid snakes (commonly called death adders), was also named after him. The collection of zoological specimens supplemented the important work he accomplished in botany, and gave him a broader perspective of the relationships among different forms of life. The unusual animals Brown observed in Australia were a continual source of fascination to those that remained at home (Fig. 10.2).

At the time, the study of plant and animal geography had not yet developed into a full-fledged discipline. The specimens Brown and others brought back provided later naturalists with information about the distribution of such unusual examples of life. When Alfred Russel Wallace traveled to the East Indies years later (1850s–1860s), he was able to draw upon his own observations as well as the work Brown and his contemporaries accomplished earlier. When Wallace went on to

---

[13] The de Jussieus, a family of botanists, were prominent in France for nearly two centuries. They included Antoine de Jussieu (1686–1758), Bernard de Jussieu (1699–1777), Joseph de Jussieu (1704–1779), and Antoine Laurent de Jussieu (1748–1836).

[14] Sir Everard Home was a brother-in-law of John Hunter (1728–1793) as well as his pupil. He was accused of plagiarizing Hunter's work and then destroying Hunter's papers to cover up evidence of plagiarism of his brother-in-law's work.

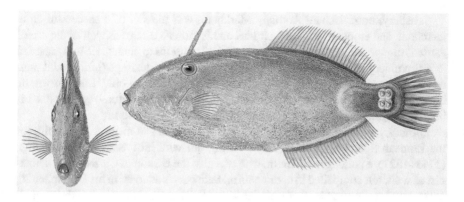

Fig. 10.2 Brown's leatherjacket (*Acanthaluteres brownii*), by Ferdinand Bauer, about 1811, ©
The Trustees of the Natural History Museum, London

develop the field of zoogeography based on his extensive observations of many
different animals from the work he conducted in the East Indies, he was able to
build upon the work of such predecessors as Brown.

The Salisbury Affair

While he was adjusting to his new role as archivist, Brown was the unfortunate
victim of some dishonesty or underhandedness. A rogue by the name of Richard
Anthony Salisbury (1761–1829) appropriated much of the information Brown had
presented in his 1809 paper, "On the Proteaceae of Jussieu," before the Linnean
Society (on January 17, 1809). Salisbury was in attendance when Brown read his
paper and he took extensive notes. Much of what Salisbury took from Brown's presentation at the Linnean Society became the crux of information that found its way
into Joseph Knight's book, *On the Cultivation of the Plants Belonging to the Natural
Order of Proteaceae*. Knight's book was published in August 1809 before Brown's
paper ("On the Proteaceae of Jussieu") could be published in *Transactions of the
Linnean Society* (1810). Thus Salisbury, through his associate Knight (George
Hibbert's gardener), was able to establish priority although much of the work contained in Knight's book was lifted from Brown's paper delivered in January 1809.
The book only contained 13 pages of cultivation techniques and over 100 pages of
taxonomic revision inspired by Salisbury's observations gained from his memoriza-
tion of terms he learned when he attended Brown's presentation.[15]

---

[15] Joseph Knight, *On the Cultivation of the Plants Belonging to the Natural Order of Proteaceae,
with their generic as well as specific characters and places where they grow wild* (London:
W. Savage, Printer, 1809). Richard Anthony Salisbury (1761–1829) (born Richard Anthony
Markham, in Leeds, England), now is regarded as making significant breakthroughs in horticulture

Salisbury, born Richard Anthony Markham, was the son of a successful cloth merchant, and enjoyed access to gardens on his father's estates as well as the use of gardens elsewhere. He had made legitimate contributions to botany. He was regarded as a prominent horticulturalist but because of his shady dealings (in financial matters as well as in science), the value of his work was often ignored. Like Brown, he rejected Linnaeus's system of classification in favor of a natural system but his botanical work was dismissed because of his intellectual dishonesty.

Initially, Salisbury was friendly with Brown's friend and mentor, J. E. Smith (of the Linnean Society), but later had a falling out with him. Samuel Goodenough (1743–1827), a colleague of Smith, who was also the Bishop of Carlisle, and a botanist as well, felt compelled to comments on Salisbury's actions. In his December 26, 1809, letter to Smith, Goodenough described Salisbury's work (in Knight's book), as a "surreptitious anticipation of Brown's paper on the New Holland plants, under the name and disguise of Mr. Hibbert's gardener."[16] This sardonic observation epitomizes how Salisbury had so tarnished his reputation that the positive work he did accomplish on his own was diminished or often ignored as a result of his indiscretions.

Salisbury had worked on Proteaceae for a considerable time—the tropical and subtropical evergreen shrubs and trees from the Southern Hemisphere Brown had observed firsthand—but because he liberally made use of Brown's observations gained from his attendance at the Linnean Society during Brown's presentation, this led to charges of plagiarism and bitterness between Brown and Salisbury. Brown did not take any action. He acted in his usual reserved manner, and allowed his friends and associates support him in this matter.

Proteaceae

Although Brown's landmark paper, "On the Proteaceae," did not initially receive priority and the recognition it deserved, it was a significant contribution to the field of botanical taxonomy.[17] This was the first published work that provided the results of Brown's Australian journey. He drew on his knowledge of nearly 200 species of Proteaceae—a family of flowering plants found to be indigenous to the Southern Hemisphere, particularly Australia and southern Africa. He also utilized his considerable microscopic skills in examining the structures of ovules and pollen. It proved

---

and botany, but his career was clouded by controversy. The adroit way he grabbed priority from Brown, as well as other things he did, highlighted his general dishonesty particularly in legal and financial matters.

[16] Letter from Samuel Goodenough (Bishop of Carlisle) to Smith, December 26, 1809, James Edward Smith, *Memoir and Correspondence of the Late Sir James Edward Smith*, ed. by Lady Pleasance Reeve Smith (London: Longman, Rees, Orne Brown, Green and Longman, 1832), vol. I, p. 587, and the letter was signed Samuel Carlisle.

[17] "On the Proteaceae of Jussieu," *Transactions of the Linnean Society* 10 (1811): 15–226.

to be a masterful work, over 200 pages long. Brown classified hundreds of species, placing them in the appropriate genera, 38 in all. The paper was mainly written in Latin, noteworthy for the time. Under each genus, he labeled categories under headings abbreviated in Latin (e.g., Char. Gen., Habitus, Desc., and sometimes obs).

Where Brown believed it was necessary, he departed from Latin and included a lengthy rationale in English, explaining the reason for placing a given species in a genus and provided his justification in assigning its name. For example, in discussing the genus *Nivenia*, he wrote, "this genus is published by Mr. Salisbury: his primary generic character does not indeed at all differ from that which he has given to Mimetes; in his account of Inflorescence, however, it is evident he understood the genus nearly as I have here proposed." He concluded, "I have therefore named it in honour of Mr. James Niven, an intelligent observer and indefatigable collector, to whom botanists are indebted for the discovery of many new species, especially in the two extensive South-African families of Erica and Proteaceae."[18]

The sheer magnitude of the task Brown accomplished was staggering. He thoroughly covered the entire Proteaceae in his analysis, including every member of the family he had observed. He concluded by remarking that those plants belonging to the Proteaceae from his "imperfect acquaintance with which, or from the unsatisfactory accounts hitherto given of them, could not with certainty be referred to any of the genera described, or, if referable to any of them, I could not with confidence propose as distinct species."[19]

What is equally remarkable besides the thoroughness of this work is that Brown was breaking with the prevailing school of plant taxonomy in Britain and the scientific establishment, in favor of the natural system of Antoine Laurent de Jussieu (1748–1836) and Augustin Pyrame de Candolle (1778–1841). Jussieu, a French botanist, was the first to publish a natural system of classification of flowering plants; his *Genera plantarum* (1789) was based on the criteria of multiple characters in classifying groups. It was an idea that originally was inspired by the work of Scottish-French naturalist, Michel Adanson (1727–1806). Adanson devised a natural system of classification and nomenclature of plants, based on all their physical characteristics, and his classification scheme centered on the taxonomic unit of family.[20]

The genius of Brown's work is that he generally applied Jussieu's classification scheme throughout the plant kingdom, a scheme that had organized the plant

---

[18] "On the Proteaceae of Jussieu," *Transactions*, pp. 133–134. Brown's naming the plant *Nivenia*, is an example of his high-mindedness, honoring someone who allowed himself, in effect, to be the instrument of Lord Salisbury's dishonesty.

[19] "On the Proteaceae of Jussieu," p. 216.

[20] Antoine Laurent de Jussieu, *Genera Plantarum: Secundum Ordines Naturales Disposita, Juxta Methodum in Horto Regio Parisiensi Exaratam, Anno 1774* ("Genera of Plants Arranged According to Their Natural Orders, Based on the Method Devised in the Royal Garden in Paris in the Year 1774"). (Parisiis: apud viduam Herissant et Theophilum Barrois, 1789). Michel Adanson's relationship with Linnaeus was amicable although they remained fierce opponents over their different taxonomic schemes.

kingdom into 15 natural groups consisting of 100 natural orders. Brown used Jussieu's orders as his starting point and then developed his own scheme similar to the one designed by Swiss botanist de Candolle. De Candolle utilized scientific criteria—drawn from his empirical studies on plant structure—in determining natural relationships in the classification of plants. Brown placed particular emphasis on the families of plants he had been able to observe firsthand in the field. Because de Candolle's method of classification allowed for a more modern interpretation of plant evolution, Darwin later embraced this system of classification. There is a suggestion that Brown may have recognized that the plants he observed and studied exhaustively were not immutable and were capable of undergoing change in structure and function. At the very least, by abandoning the traditional taxonomic scheme of Linnaeus, so popular in the English-speaking world, Brown demonstrated that he was willing to break with convention and was open to new ideas. But he did not speculate on such matters early in his career. Most certainly, Brown did not suggest that species changed, nor did he consider a mechanism for species change.

Prodromus Florae Novae Hollandiae

Brown's paper on Proteaceae was comprehensive and accomplished with great precision, so much so that there were few reasons to quibble with his analysis. The breathtaking scope with which Brown analyzed the Proteaceae gave him support in the controversy with Salisbury (who had already become quite discredited). And when Brown made use of Salisbury's work as well as Niven's and other observers such as Knight, he gave due credit to them, even to go as far as naming genera after them. Brown's nature was not combative in anyway, and he exhibited great generosity considering the manner with which he had been treated. The result was that his work enhanced his reputation as an important naturalist. His daring work, breaking with Linnaeus's artificial system, set the stage for his book, *Prodromus Florae Novae Hollandiae*, published in 1810, including some of the information found in his earlier paper.[21] *Florae Novae* was intended to include the genera and species of all the plants he determined to be indigenous to Australia (New Holland). He wrote to Dawson Turner in February 1810, "The first part of my Prodomus … is in considerable forwardness … it is a odd kind of publication," which he believed to be unique in that he could not think of anything as "applicable as Prodromus."[22]

To establish priority over the French, it was published and distributed in March 1810. Brown bore the cost for its printing and he presented 24 copies to learned societies and the prominent botanists of the day. He attempted to sell the remainder of the 250 published copies but was only successful in selling 24 more copies. At

---

[21] Brown, *Prodromus Florae Novae Hollandiae et Insulae Van Diemen* (Prodromus of the Flora of New Holland and Van Diemen's Land) (London: Richard Taylor & Son, 1810).

[22] Letter from Brown to Dawson Turner, February 1810, *British Library, Add, MSS* 32439 ff. 301–302.

the time, his intention was to publish at least one more volume but he only published the first one consisting of 450 pages despite the fact that there was enough material to publish a second volume. He may have been discouraged by the first volume's lack of commercial success. In addition, Brown had little support from Banks in publishing a second volume. Brown, however, continued to work on the second volume until around 1817. The second volume would have covered the Leguminosae, Myrtaceae, Compositae, as well as other families of plants. Happily, Brown incorporated much of this material in various papers and books describing his exploration in natural history, such as his appendix to Flinders's *A Voyage to Terra Australis* (1814).[23]

Prodromus Florae Novae Hollandiae remains a significant work; its scope encompasses much of Australian botany. It begins with the more primitive vascular plants, Filices (ferns), and related groups, using the natural system initially employed in his 1809 paper. He then went on to describe monocotyledons, particularly the Gramineae, and finished with 37 families of dicotyledons, including Proteaceae, Scophulariaceae, Apocynaceae, and Asclepiadaceae, and the cycads. It covered 464 genera and approximately 2000 species. The *Prodromus* bears much of the influence of de Jussieu and de Candolle but Brown broke new ground by utilizing a different sequence of the plant orders. This was because he found it difficult to place all the Australian plants he studied into the smaller groups that de Jussieu and de Candolle had constructed and as a result he developed several "new natural orders" which he indicated that he had "not included" initially. Brown's language was lucid, and the work displayed his outstanding grasp of taxonomy and his masterful powers of observation. He modified some of the families of de Jussieu and de Candolle, which he had previously begun in his 1809 paper. It can be argued that Brown "presided over" the "disintegration of the Linnean system" and that his genius in doing so made him perhaps one of the most important figures in the whole history of British botany.

Brown continued to receive fresh specimens from many regions of the world because, during his tenure as Banks's librarian and subsequent position as the first keeper of the Botanical Department of the British Museum, he had unique access to material gathered from distant places. Brown sat "spider-like at the centre of the botanical web," with material coming to him from collections from the entire world.[24]

---

[23] Gay Hatfield, in her monograph "Robert Brown A.M. F.R.S.E. (1773–1853)," included in *Scottish Men of Science* (Edinburgh: University of Edinburgh Press, 1981), indicated that because it received a "critical reception" in the Edinburgh Review, not over its content but its "poor quality of Latin," Brown withdrew it from circulation. She did not provide any substantiation for this claim but suggested that this experience caused Brown to be "unduly cautious in the wording of everything he subsequently wrote," p. 2.

[24] John Gilmour made this claim in his work, *British Botanists*, indicating that Brown "presided over" the "disintegration of the Linnean system" and his genius in doing so made him "perhaps the greatest figure in the whole history of British botany" (London: William Collins, 1944), p. 33. Gilmour added that during his tenure as Banks's librarian and subsequent position as the first keeper of the Botanical Department of the British Museum, Brown presided "spider-like at the

Botanicorum Facile Princeps

Brown's acceptance of the natural system of classification led to a better understanding of the reproduction, anatomy, characteristics, geographical distribution, and relationships of different groups in the plant kingdom. Because this led to a wider acceptance of the natural systems of classification, it represented an important break with the artificial system of Linnaeus, and confirmed Brown's status as one of the most significant botanists in the nineteenth century. The widespread approval of Brown's contributions and the acclaim he received in the botanical sciences were epitomized by the title von Humboldt bestowed upon him, "Botanicorum facile Princeps."[25]

In spite of the good reception that publication of *Prodromus* received, its cost and limited sales eventually inhibited Brown from publishing the second volume. Because the published volume was relatively slender for works of this kind and had a limited printing of only 250 copies, it was rather expensive at the time, selling for 18 shillings. In addition, it was not printed particularly well because the quality of the paper was relatively poor. There were no illustrations and there was no index.[26] Curiously, it appeared that Banks was not particularly interested in Brown's future efforts because there were no plans for further support.

Brown possibly was discouraged and may have toyed with the idea of emigrating, perhaps to Australia. However, there is no strong evidence that he seriously contemplated such a course of action and this became immaterial at any rate because in October of 1810, Banks's trusted librarian, Jonas Dryander (1748–1810), died. Brown's role had now changed. In Banks's eyes, Brown became indispensable to him and his work because of Dryander's death. He appointed Brown to be his librarian and assumed the responsibility of editing William Townsend Aiton's (1731–1793) *Hortus Kewensis* (A Catalogue of the Plants Cultivated in the Royal Botanic Garden at Kew), the 1789 catalogue of all the plant species that were being cultivated at Kew representing most of the plant species in England. The *Hortus Kewensis* included information regarding the country of origin, and identified the person who introduced the species into England.

Banks recommended Brown to undertake this task because William Aiton's son requested that the second edition of his father's work should utilize Brown's nomenclature in the revision. The acute microscopic skills Brown employed in observing the structures of ovules and stamens (pollen) as well as his findings in plant development further enhanced his status as a preeminent botanist at the time by his contemporaries. Much of the pioneering work on Australian plants that Brown intended

---

centre of the botanical web." All material, it seemed, came to him from plants that were collected from the entire world.

[25] Later on, in 1859, Joseph Hooker of Kew described it as the "greatest botanical work that has ever appeared."

[26] However, although the paper is rather coarse in texture, not glossy, and there is some foxing, the copy in the N.Y. Public Library remains in remarkably good condition.

for the second volume of *Prodromus*, such as his work on Leguminosae, was included in Brown's edited edition of *Hortus Kewensis*. As a consequence of accomplishing all this work, Brown was awarded a stipend of £200 per annum, adding to the £100 annual income he received from the Linnean Society for eight hours of work a week he performed there. This provided him with further financial security, always an important consideration as far as he was concerned. In addition, Brown was elected a Fellow of the Royal Society in 1811, further affirming his status as a first-rate scientist.[27]

Despite Brown's contributions to science in general as well as further enhancing Banks's considerable prestige, Banks remained quite preoccupied over the death of his trusted aide, Dryander, so much so that he did not seem to value the work of his loyal and able assistant (Brown). Banks wrote to Everard Home (1756–1832)—who was also his personal physician and a naturalist and an expert on anatomy—on October 22, 1810: "I was so stunned by the unlooked Blow … I have lost my right hand, & can never hope to provide any thing as a Substitute that can at all make amends to me … I always hoped that Dryander would outlive me: he was younger & less afflicted with disease than myself."[28] It is obvious that at the time, Banks did not wholeheartedly believe that Brown would be able to fill the gap left by Dryander's death. Brown's close associate, J. E. Smith, did not agree with Banks. Smith wrote to his friend Samuel Goodenough (the Bishop of Carlisle) and confided, "I find Mr. Brown is in Dryander's place at Soho Square;—his manner will be more *suaviter* (gentler) but not less *fortiter* (stronger) with coxcombs and blockheads."[29]

In hindsight, Banks's relatively blasé or cavalier attitude about the value of Brown's work seems strange in view of Brown's great contributions in botany and explorations in natural history. Brown did everything that was asked of him, with characteristic modesty and without complaint. It appears that talented naturalists, such as Brown, Park, Caley, and other contemporaries, were viewed by the British scientific establishment and the Admiralty as pawns to be used on behalf of Britain's colonial ambitions, i.e., "to help plant the British flag in distant lands," and further develop markets for British goods. The scientific work they accomplished seemed secondary to establishing primacy over France and other colonial rivals. The French actually invested more in their efforts in natural history; their ships were in much better shape, quite seaworthy and well manned. That stands in sharp contrast to the conditions aboard the *Investigator* for example. Yet due to the ability of men such as Brown, their accomplishments in natural history and exploration stood out.

---

[27] In spite of this recognition, he may have given some consideration to resuming his medical studies at the time. This may have been more of a commentary of the insecurity faced by naturalists like Brown who did not come from privileged circumstances.

[28] Letter from Banks to Everard Home, October 22, 1810. *The Indian and Pacific Correspondence of Banks*, vol. 6, Letter # 1942, p. 48.

[29] Letter from Smith to Bishop of Carlisle, December 3, 1810, James Edward Smith, *Memoir and Correspondence of the Late Sir James Edward Smith*, ed. by Lady Pleasance Reeve Smith (London: Longman, Rees, Orne Brown, Green and Longman, 1832), vol. I, p. 592.

The treatment of Matthew Flinders is also noteworthy. He remained imprisoned by the French until 1810 despite frequent letters from his wife Ann to Banks and other officials. Despite his failing health, it seems that efforts to enable him to go free were less than vigorous. When Flinders finally was released in June 1810, he wrote to Banks a day after he arrived in England (from London), exclaiming to Sir Joseph that "I have the happiness to inform you of my arrival in England yesterday morning, and in town this evening. The circumstances of my release are as extraordinary as were those of my imprisonment, but they are too long for me to detail at this moment, suffice it to say, that I quitted the I.[sle] of France on June 13, and the Cape on Aug. 20." He went on to report that he had seen the First Lord of the Admiralty, Charles Philip Yorke, concerning the backdating of his promotion that was promised to him by Lord Spencer, and he would visit Banks soon and detail his detention and ultimate release. Thus ended a rather shameful episode but Flinders's treatment must have made an impression on naturalists such as Brown.[30]

In spite of this, Brown continued to draw admiration and respect from his colleagues, and to him this seemed to be sufficient. Brown was happy in gaining an additional source of income as Banks's librarian, and was particularly pleased that he no longer had to go back to the military to earn his living. He was a full-fledged professional scientist, one of the few men who enjoyed that status in the scientific world at the time, and his salary earned from his scientific work enabled him to adequately support himself.

---

[30] Letter from Flinders to Banks, October 25, 1810, written from the Norfolk Hotel in London, *B.M. Add. MS.* 32439. 332. Also in *The Indian and Pacific Correspondence of Banks,* vol. 8, Letter # 5, p. 4.

Chapter 11
Banks's Librarian

Contents

Brown's appointment as Banks's librarian meant that he was now in charge of Sir Joseph's entire library as well as his collections. Brown's presence, his methodical work habits, and his expertise were indispensable at this critical moment because Banks began to suffer increasingly from a variety of health problems, particularly gout. Through Brown's diligence, Banks's library and collections remained the center of British botany. Brown's April 6, 1813, letter to J. E. Smith reported, "Sir Joseph Banks is slowly recovering from his very tedious fit of gout which has so much reduced his strength that I fear it will be yet some time before we have him again in the Library." This remained very much a concern even when Banks's health improved. In Brown's December 1, 1813, letter to Smith—where Brown indicated that Smith's genus *Poiretia* was antedated by Étienne Pierre Ventenet's in *Mem. Inst. Fr.*, 1807, 4—he informed Smith that "Sir Joseph Banks at present in excellent health." Several years later (1818), in an August 14th note to Smith in which Brown informed Smith that he sent a "box" to him with books, Brown reported that his own health continued to be "very indifferent" but "Sir Joseph Banks is quite well," adding that he "comes to town twice a week."[1]

---

[1] Brown's letters to Smith, Linnean Society; Brown's letter to Smith, April 6, 1813, 2:156; Brown's letter to Smith, December 1, 1813; Brown's letter to Smith, August 14, 1818; and Smith's letter to Brown are in the *British Library, Add. MSS* 32430, f. 274, 279, 285, 303, and a copy is in the Botany Library of the Natural History Museum.

© The Author(s), under exclusive license to Springer Nature Switzerland AG 2021
J. Schwartz, *Robert Brown and Mungo Park*, Memoirs of The New York
Botanical Garden 122, https://doi.org/10.1007/978-3-030-74859-3_11

Botany of Terra Australis

One of the important works produced by Brown during the decade following Dryander's death in 1810 was Brown's memoir inserted as a botanical appendix to Flinders's *A Voyage to Terra Australis*, "General Remarks, Geographical and Systematical, on the Botany of Terra Australis" (1814). Brown's "General Remarks" contained the material he originally intended to publish in a second volume of *Prodromus*. In the botanical index, Brown included and carefully arranged the plants from the Leguminosae (the pea family whose members contain pods which serve as its fruit), Myrtaceae (dicotyledonous plants from the Myrtle family), and Compositae (including Chrysanthemum, Endive, and Chicory genera). Also included in this list were the families he established, Pittosporaceae (cheesewoods and lemonwoods and other tropical flowering plants), Cunoniaceae (wild alder for example), Rhizophoraceae (tropical or subtropical flowering plants, for example mangroves), Celastraceae (staff vines or bittersweet family), Haloragaceae (the water milfoil family), and Stackhousiaceae (which has no common name, except it is generally referred to as the Stackhousia family). These plant families were now consistent with the natural system Brown developed on his own, the result of his modification of the French or European system he found quite superior to the old one it replaced. The evidence drawn from his collections and observations of Australian plants supported it.

In describing the plants from these groups, Brown considered their geographical distribution as well as their anatomical and physiological adaptations. Brown noted that their transport was facilitated to different regions (by water or air), and that some of them had traveled (migrated) to Africa. He also calculated that he had studied approximately 4200 species indigenous to Australia, and further estimated the number of dicotyledons as three times the number of monocotyledons.[2] Brown's appendix contained so much information concerning the distribution of plants he observed, collected, and studied in detail that it marked a significant contribution to the study of plant geography. Because plant geography was a discipline Alexander von Humboldt helped to develop and remained keenly interested in, Brown's contributions in this area added to von Humboldt's admiration for Brown.

---

[2] Robert Brown, "General Remarks, Geographical and Systematical, on the Botany of Terra Australis," in Mathew Flinders, *A Voyage to Terra Australis; Undertaken for the Purpose of Completing the Discovery of that Vast Country, and Prosecuted in the Years 1801, 1802 and 1803, in His Majesty's Ship the* Investigator. 2 vols & Atlas (London: G. & W. Nicol, 1814). [Reprinted in *The Miscellaneous Botanical Works of Robert Brown* John J. Bennett ed. (London: Published by Ray Society for R. Hardwicke, 1866–68)].

Reproductive Structures in Mosses

Brown's work in plant histology and morphology that he undertook soon after his return from Australia continued to use the same careful methods of investigation and observation he previously employed. The meticulous way he went about looking at the material he and others gathered enabled him to integrate the structural details of the different plant species he examined, and then use his observations to support the taxonomic scheme he was developing. An example of such thorough study was his 1809 paper—read before the Linnean Society on June 20, 1809, and published in 1811—"Some Observations of the Parts of Fructification in Mosses; with Characters and Descriptions of Two New Genera of that Order"—where Brown focused on the reproductive parts of mosses.

Brown's observations on the "fructification in mosses" critically examined the previous findings of Palisot de Beauvois (Ambroise Marie Francois Joseph Palisot, Baron de Beauvois (1752–1820), who, in Brown's words, regarded "the capsule of Mosses ... as the containing organ of both sexes ... the granules which Hedwig supposes to be seeds, he regards as pollen; the real seed according to him being imbedded in the substance of the body which occupies the centre of the capsule, and to which botanists have given the name of *columnula* or *columella*."[3]

Johann Hedwig (1730–1799) was the botanist Brown was referring to in his critique of Baron de Beauvois's work. Hedwig was an expert on mosses and their reproduction. He had identified the female reproductive organs (now called archegonia) and the male reproductive organs (now called antheridia). Brown indicated that these structures were separate, explaining that they produced the male and female gametes independently. Brown concluded that they did not follow Beauvois's model of both sexes residing in a single "capsule." He reported, "the conversations I have had with my ingenious and accurate friend Mr. Francis [Franz] Bauer, as well as observations of my own, I am disposed to believe that considerable diversities may exist .... [and] militates against M. Beauvois's theory." Brown went on to describe "two new genera," and he devoted most of the paper in listing the separate species belonging to each genus. He thoroughly enumerated the characteristics of each group in Latin, as a rationale in altering the previous taxonomic scheme.[4]

Brown corrected the erroneous assumption of de Beauvois and Hedwig that mosses were seed-producing plants. His earlier association with Dickson and Dickson's interest and knowledge about cryptogams may have assisted him in understanding that they were quite different from flowering plants. This allowed him to gain insight into the nature of mosses and similarly related plants, i.e., the life cycles of these plants and how they reproduced.

---

[3] Brown, "Some Observations of the Parts of Fructification in Mosses; with Characters and Descriptions of Two New Genera of that Order," *The Transactions of the Linnean Society of London* (1811) 10: 312–324, 315.

[4] Brown, "Some Observations of the Parts of Fructification in Mosses," *The Transactions of the Linnean Society of London* (1811) 10: 312–324, 316–324.

Caley's Assessment of Brown's Work

Brown informed Banks about his progress, often by post when Banks was at his family home in Revesby Abbey, Lincolnshire. For example, he wrote to Banks—on October 3, 1815—that he was leaving for Manchester and would return in 10 days, informing him that "George Caley was already there." He indicated that he would meet Dr. George Young and "other persons of St. Vincent, whither Caley is going." The letter concerned Banks's efforts to help Caley secure a position.[5] In spite of Caley's numerous complaints and sometimes acerbic comments, Banks showed unusual patience toward Caley. At that time, Banks was recommending Caley to be the curator of the St. Vincent Botanical Gardens, in the British West Indies.

Brown also exhibited a good deal of patience and forbearance toward Caley because Caley sometimes made Brown the target of complaints to Banks. In one such letter to Banks, dated September 27, 1807, in which he railed on bitterly about conditions in Australia—"the place is miserable beyond description"—he wrote, "The garden as Mr Brown left [it] wants considerable repair which I shewed him. He pretended he [k]new what was wanted in this line & speaking in an overbearing & piercing tone which hurt my feelings ... and made my blood to rise & we had a few words. He afterwards told me I was a great expence [sic] to Government and that I was humbugging him."[6]

Caley wrote a lengthy letter to Banks a year later again while he was still in Australia. On July 7, 1808, he wrote, "I cannot how but remark how lucky Mess. Brown and Bauer were in visiting this country; though I by no means admire the manner in which they left it. It has been reported that Mr Brown's collection was spoiled; but as I have not heard from you, nor from him, I now conclude it must have been a fabricated rumour. I have something to say of Mr Brown that by arriving & collecting so much, in so short a time, a deal of his work must not be well done."[7]

In spite of the negative comments that Caley made to Banks, and additionally his criticism of Brown, both Banks and Brown valued Caley's skills as a naturalist so they put up with his acerbic remarks. They understood that if they could overlook Caley's temperament, his work would be very useful. Brown showed his generous nature when on one particular occasion, he referred to Caley as "botanicus peritus et accuratus" (skillful and accurate botanist).[8] In any event, Caley's comments about Brown represented a minority view because Brown's work continued to draw admiration. One such example, but certainly not an isolated one, was what Swedish

---

[5]Letter from Brown to Banks, Oct. 15, 1815, *B.L., Add. MS.* 33,982. 90–91, and copy in *D.T.C.*19:198–200.

[6]Letter from Caley to Banks, Sept. 27, 1807, *B.L.* (N.H. B.C. 140), and in *The Indian and Pacific Correspondence of Banks,* vol. 7, Letter # 157, pp. 279–281, p. 281.

[7]Letter from Caley to Banks, July 7, 1808, *D.T.C.* 17: ff. 168–198, and in *The Indian and Pacific Correspondence of Banks,* vol. 7, Letter # 186, pp. 380–394, p. 392.

[8]Joan Betty Webb, *George Caley, Nineteenth Century Naturalist, A biography* (Chipping Norton: S. Beatty & Sons, 1995), p. xi.

botanist Olof Swartz wrote to J. E. Smith in 1817, telling him "Mr. Robert Brown has also favoured me some months ago. It is with great impatience that I wait the sequel of his admirable Prodromus."[9]

Brown's Association with Smith and Other Naturalists

Brown continued his close relationship (and friendship) with Smith, informing him about general matters such as the state of Banks's health and his own well-being, and also botanical concerns. In 1810, Smith named a genus of the small herbs, blue pincushion (or native cornflower), *Brunonia australis*, after Brown. Brown had collected the plant in Victoria, Australia, and described it in the *Prodromus*. The interesting thing about the plant was that it was unigeneric or monotypic, meaning it belonged to a family, Brunoniaceae, that consisted of only one genus (Figs. 11.1 and 11.2).

However, Smith's generous act created a problem. *Brownea* had already been used *to* honor a Jamaican naturalist, Patrick Browne. Smith then changed the name of the genus from *Brownea* to *Brunonia* and presented the change of name before the Linnean Society on February 6, 1810, thereby honoring both men. Smith believed that this was a satisfactory accommodation. The difficulty arose because Brown's information about the genus and family named in his honor was first published in the *Prodromus*, later in 1810. Therefore, Brown actually was the first to refer to this newly named genus in print because Smith's proposed change of name was not published until 1811 (in volume 10 of *Transactions of the Linnean Society*). Therefore, it appeared that Brown had published his own honorific. This episode proved to be rather awkward at best but there is no question that it was a most deserving honor. It demonstrated the regard people like Smith had for Brown and his accomplishments.[10]

Brown received more specimens from other naturalists and colleagues such as Caley and Macleay in addition to collectors mainly from Great Britain, e.g., William Borrer (1781–1862), an expert on British botany and collaborator with Dawson Turner on a work on lichens; Charles Lyell—Lyell sent him bryophytes collected on a trip to Mount Snowden in Wales—Thomas Taylor (1775–1848), a specialist whose work focused on cryptogams, specifically bryophytes (mosses), and who was

---

[9] Letter from Olof Swartz to James E. Smith, November 19, 1817; James Edward Smith, *Memoir and Correspondence of the Late Sir James Edward Smith*, ed. by Lady Pleasance Reeve Smith (London: Longman, Rees, Orne Brown, Green and Longman, 1832), vol. I, p. 497. Olof Swartz (1760–1818) was a Swedish botanist and a student of Linnaeus. A taxonomist, Swartz studied pteridophytes, a group of vascular plants, mainly ferns, that do not have seeds.

[10] James E. Smith, "An Account of a Genus of New Holland Plants named *Brunonia*." *Transaction of the Linnean Society of London* 10: 365–370. Smith's paper contained Ferdinand Bauer's illustrations of both species of *Brunonia*. The episode is described in Mabberley's *Jupiter Botanicus*, pp. 165–166, and Helen Hewson's *Brunonia australis: Robert Brown and his contribution to the Botany of Victoria* (Canberra, Australia: Centre for Plant Diversity Research, 2002), p. 2.

Fig. 11.1 *Brunonia australis* (Blue Pincushion) by Sarah Anne Drake from *Edwards Botanical Register*, New Series, Volume IX (1836), New York Botanical Gardens, The LuEsther T Mertz Library

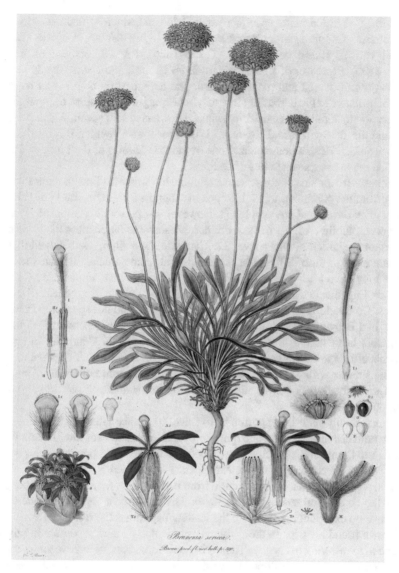

Fig. 11.2 *Brunonia australis* (Blue Pincushion), State Library Victoria

an associate of William Jackson Hooker; and finally Friedrich Sellow (1789–1831), a German-born botanist whose specialty was Brazilian flora. But perhaps the most productive correspondence he had was with the French naturalists and their disciples. The Napoleonic Wars were ending so better communication between British and French naturalists was now possible.

Brown forwarded many specimens that Caley and Macleay had sent him from Australia to Antoine Laurent de Jussieu (1748–1836), Ambroise Marie Francois Joseph Palisot, Baron de Beauvois (1752–1820), René Louiche Desfontaines (1750–1833), Professor of Botany at Jardin des Plantes, and Louis Claude Richard (1754–1821), botanist and plant illustrator and an expert on conifers as well as a chief contributor to the herbarium in Paris, in keeping with the spirit of international cooperation fostered years earlier by Banks. Brown also sent specimens to Augustin de Candolle in Switzerland, Swartz in Thunberg, and Aylmer Bourke Lambert (1761–1842), a British botanist, and one of the first fellows of the Linnean Society who had made extensive studies of conifers.

Because Brown's most constructive relationship was with French naturalists and others from the continent, this reinforced his support for the natural system of taxonomy they developed and employed. However, his personal relationship with the elder de Candolle (Augustin) was strained because the latter most likely resented Brown and regarded him as a rival. De Candolle's son Alphonse believed that this was due more to their different backgrounds and temperaments. Brown's reserved nature may have been a contributory factor as well. When Augustin de Candolle came to England in January of 1816, the visit was marked by strained circumstances. De Candolle did not speak English and Banks did not speak any French so Banks had Brown show him the library and the herbarium. Brown tried to be a hospitable host, taking him to dinner and to Kew, but de Candolle, in spite of Brown's efforts to be congenial, did not reciprocate. De Candolle seemed to be on friendlier terms with Salisbury, who was by nature a more convivial or outgoing sort. Curiously, Salisbury was still permitted to use Banks's library in spite of his semi-outcast status.[11]

Brown finally had the opportunity to travel to France later in 1816, beginning what would become a regular and quite enjoyable experience for him. Joseph Woods (1776–1864), an English architect who became interested in botany—who eventually became a Fellow of the Linnean Society—preceded Brown's initial visit to France, carrying a letter of introduction from Brown to Professor Desfontaines. This proved to be a useful stratagem for Brown. His subsequent visits to France helped him develop excellent rapport with the leading naturalists there (Fig. 11.3).

Brown found his trips to the continent, particularly France, exhilarating as well as a balm in restoring his health. These visits reinforced his innovative views regarding taxonomy, and he developed friendships with leading French botanists and taxonomists such as Louis Claude Richard; Professor Desfontaines, with whom he had corresponded but only met him for the first time during his early trip to France; Palisot; Nicaise Auguste Desvaux (1784–1856), director of the Botanical Gardens in Angers; Jacques Labillardière (1755–1834), an expert on Australian flora; Jules Paul Benjamin Delessert, Baron (1773–1858), a wealthy banker, investor, and amateur botanist; and André Thouin (1746–1824), a student of Bernard de Jussieu.

---

[11] Mabberley discusses at length why the elder de Candolle did not get on with Brown in *Jupiter Botanicus*, pp. 197–198.

Fig. 11.3 Sir Joseph Banks's study in his house at 32 Soho Square, London, by Francis Boott, 1820. Inscription on the back reads "Museum and Library of Sir Joseph Banks, Bart. 17 Dean Street, Soho: The residence from 1820 to 1858 of his librarian, Robert Brown, …, who died there, June 10, 1858, in the 85th year of his age." Library and Archives Collection at the Natural History Museum, London, 13204, © Trustees of the Natural History Museum, London

Brown brought with him a good deal of botanical material and presented samples of specimens to the Paris Museum as well. He enjoyed the opportunity to discuss botanical taxonomy firsthand with kindred spirits, those who had views that were very compatible with his own. They not only welcomed Brown's support but also relished the opportunity to examine the material brought with him. Later on in the year after his return from Paris, 1816 (December 14, 1816), Brown wrote to Smith that his "trip to Paris was altogether very agreeable," indicating that the "continent" was "quite new to me."[12]

Brown began to take up his work on the second volume of *Prodromus* again. During this pivotal year, 1816, he also met Alexander von Humboldt at Kew during Von Humboldt's visit to the Royal Society. Von Humboldt was eager to learn about the Australian flora Brown had described and collected, and the much-revered scientist further encouraged Brown in his efforts to adapt the natural taxonomic system buttressed by the vast evidence he had accumulated. Von Humboldt's close

---

[12] Brown's letter to Smith, December 14, 1816 (Letter # 10), the *Linnean Society of London*.

associate Aimé Bonpland (1773–1858) already had visited Britain and during his stay there, Brown shared information with him. In turn, when Brown and Von Humboldt met, the latter also supplied Brown with valuable information regarding the geographical distribution of plants in the Western Hemisphere.

Brown enjoyed meeting with such illustrious colleagues. But the general state of his health was a nagging concern to him. Several years later (1818), Brown commented to Smith that his "health continues to be very indifferent that I feel it almost necessary to leave London for a few weeks" and he may "go to France," promising to "set off in a few days." There was no specific complaint so one can only gather it was due to the effects of his rigorous Australian journey more than a decade before and the rigorous schedule of work he had devoted himself to. He was no longer a young man. In the same letter, Brown informed Smith about a box he was sending him, "containing books," and was busily engaged in the "printing of the transactions" of the Linnean.[13]

The Herbarium of Professor Christian Smith

Brown produced other significant findings in his botanical appendix of Flinders's *Voyage*. He found similarities between the plants found in the "horn of Africa" (present-day Ethiopia and Somalia, then referred to as Abyssinia), with some of those indigenous to South Africa. He had the opportunity to examine the work of other naturalists. Christian Smith (1785–1816) was one such example. He was a Norwegian physician, naturalist, economist, and explorer, and was asked by the Royal Society to serve as a naturalist to Captain Kingston Tuckey. The object of the expedition was to compare vegetation between the regions of the Congo River and the Niger River basins in western and central Africa. The parallel between Christian Smith and Mungo Park was inescapable. Some of the regions explored by Smith and Tuckey were in the same territory previously explored by Park and this journey also ended in disaster; that is, there were hostile native people, various tropical afflictions, and a lack of food. Christian Smith died—probably of yellow fever—but not before his diary and botanical specimens were shipped back to London. 620 species were in the collections assembled by Smith, with 250 that were heretofore undiscovered. But a terrible cost was paid because 18 of the 56 members of the expedition died, including Captain Tuckey in addition to all the scientists who had accompanied him in the party.[14]

Brown was asked by Banks to examine Christian Smith's herbarium. He studied Smith's material very carefully, classifying the plants in this collection in accordance with the natural system he adopted and modified, very much in same fashion

---

[13] Brown's letter to Smith, August 14, 1818 (Letter # 13), the *Linnean Society of London*.

[14] James Hingston Tuckey and Christen Smith. *Narrative of an Expedition to explore the River Zaire usually called the Congo in South Africa, in 1816* (London: John Murray, 1818).

employed in his botanical appendix to Flinders's *Voyage*. He paid careful attention to the geographical distribution of the plant species, and considered their seed dispersal—how their structural and physiological adaptations helped their transport—speculating that they were imported into central Equatorial Africa as well as the "Horn." For example, he noted how their seeds resisted the action of saltwater, and how they passed through the digestive systems of birds without any alteration. He also focused on the different number of species belonging to dicotyledons and monocotyledons, and was struck by the greater number of dicotyledonous species in Christian Smith's herbarium.[15]

Henry Salt's Collection from Abyssinia

Brown was predisposed to see the similarity between the plants from South Africa with the species from Abyssinia (Ethiopian region of Africa) when he examined Christian Smith's herbarium. A few years earlier, Banks had asked him to look at the collection of plants assembled by Henry Salt (1780–1827). Salt was sent by the British Government to explore and collect, and he assembled an impressive collection of antiquities as well as a herbarium from "Abyssinia" which he sent to Banks. Salt's account of his journey, *A Voyage to Abyssinia*, was published in 1814, based on his travels conducted from 1809 to 1810.[16] Brown's analysis of Salt's findings was published as an appendix to Salt's account of his voyage, which represented the first description of African plant species. Initially, Salt's collection was classified according to the Linnean system but this was a prelude to the larger task of analyzing subsequent collections and reclassifying them. During his examination of Salt's material, Brown discovered a strong similarity between the flora of Abyssinia and that of southern Africa, which reinforced his ideas concerning seed dispersal.

---

[15] Brown, "Observation, Systematical and Geographical, on the Herbarium collected by Professor Christian Smith in the Vicinity of the Congo, during the expedition to explore that river, under the command of Captain Tuckey, in the year 1816," in *The Miscellaneous Botanical Works of Robert Brown*, Volume I, edited by John J. Bennett (London: published for the Ray Society by R. Hardwicke, 1866–68), pp. 97–175.

[16] The full title of Salt's work was *A Voyage to Abyssinia, and Travels into the Interior of That Country, Executed under the orders of the British Government, in the years 1809 and 1810*. (London: Printed by F. C. and J. Rivington, by W. Bulmer and Co.1814). The Appendix containing Brown's analysis is on pp. lxiii-lxv. It was reprinted with the title, "List of new and rare Plants, collected in Abyssinia during the years 1805 and 1810, arranged according to the Linnean System," and is in *The Miscellaneous Botanical Works of Robert Brown*, Volume I, edited by John J. Bennett (London: published for the Ray Society by R. Hardwicke, 1866–1868), pp. 91–95.

The Largest Flower in the World

Somewhat indirectly, Brown found himself involved in an interesting discovery during this period. Sir Thomas Stamford Bingley Raffles (1781–1826), the Governor of the East India Company's holdings in Sumatra, asked Banks to "use his influence" with the East India Company to select a well-qualified naturalist to help explore the region. Raffles, the founder of Singapore, was a naturalist himself and believed that Banks's input would be useful. There was some discussion and after William Hooker decided not to go to Sumatra on the advice of his father-in-law, Dawson Turner. Turner or possibly even Brown recommended that Dr. Joseph Arnold (1782–1818)—a naval surgeon and friend of Turner as well as a man who was well regarded by Brown and Banks—go as a naturalist. Arnold was a bit nervous about this undertaking because his experiences on previous voyages were marked by a fire aboard the *Indefatigable* where he lost many of his journals and insect specimens on one occasion, and had a difficult passage back from Australia with Captain Bligh after the "Rum Rebellion" on another. Because Arnold had been elected to the Linnean Society in 1815, he anticipated settling in England but those plans were shelved when he became a naturalist for the expedition.[17]

Arnold moved back to England in 1817 to prepare himself for the task. He sought out Banks's assistance and the latter placed him in Brown's capable hands. Brown suggested that Arnold should live nearby Banks's library while preparing for the trip and Brown recommended books that would help in his preparations. Arnold liberally used Banks's library, and found the drawings assembled by the Scottish botanist and surgeon, William Roxburgh (1751–1815), particularly useful. Roxburgh, like Brown and Park, received his education at Edinburgh University, although he was there somewhat earlier than when they attended. Roxburgh went to India to explore the indigenous flora. He was so successful in his work that he became known as the "father of Indian Botany." The magnificent drawings made by Indian artists who assisted him served as an important resource for those wishing to travel to that region.[18]

In 1818, Arnold traveled to Sumatra with Raffles and his wife, Lady Raffles, who possessed some artistic ability as well. On one excursion on May 19, 1818, at Pulau Lebar on the Manna River, they discovered an unusually large flower that was part of a parasitic plant living on the vine, *Tetrastigma*. Because of its parasitism, it was able to thrive without developing any evident roots or leaves. During the time he was examining this unusual plant, Arnold developed a serious fever—probably

---

[17] Banks. *The Indian and Pacific Correspondence of Sir Joseph Banks, 1768–1820*, Volume 8, Letters 1810–1821, ed. Neil Chambers (London: Pickering and Chatto Ltd., 2014). Also see John Bastin, "Dr. Joseph Arnold and the Discovery of *Rafflesia arnoldii* in West Sumatra in 1818," *Journal of the Society for the Bibliography of Natural History* (1973) 6(5): 305–372, 313–314.

[18] See Tim Robinson's *William Roxburgh. The Founding Father of Indian Botany* (Chichester, England: Phillimore, in association with the Royal Botanic Garden, Edinburgh, 2008).

malaria—and died, and Lady Raffles finished the drawing begun by him of what was regarded as the largest flower in the world.[19]

The drawing and the plant's preserved material were sent to Banks who passed it on to Brown as well as Franz Bauer, then serving as the resident artist at Kew. Brown carefully examined the flower in detail, and reported his findings before the Linnean Society (read on June 20, 1820, and published in 1822). Brown's paper devoted considerable space to the plant's peculiar anatomical characteristics and its taxonomy.[20]

Brown bestowed the scientific name, *Rafflesia arnoldii* on this plant (commonly called corpse lily today), in his paper on the genus. He initially wanted to name the genus *Arnoldia*, in honor of the naturalist who lost his life during the exploration.[21] The episode is instructive in illustrating how Brown became involved with the work of other naturalists. He frequently was called on to examine the material collected by other naturalists. Although this work was time consuming and often tedious, Brown found such efforts useful in refining his classification of the unusual plants sent to him and reinforcing his own ideas regarding the fine structure of plants and the broad taxonomic scheme he was developing (Figs. 11.4 and 11.5).

In similar fashion, Brown spent time studying the botanical material from Captain John Ross's exploration to Baffin Bay in 1818 which resulted in his "List of Plants collected by the Officers, &c., in Captain Ross's voyage, on the coasts of Baffin's Bay," an 1818 paper listing the plants discovered during Captain Ross's voyage to Baffin's Bay. The notable finding in this work was Brown's explanation for the appearance of "red snow"; it was caused by a species of alga.[22]

A few years later, Brown had the opportunity to examine plants that were collected by another explorer-naturalist, Captain William Scoresby (1789–1857).

---

[19] John Bastin, "Sir Stamford Raffles and the Study of Natural History in Penang, Singapore and Indonesia," *Journal of the Malaysian Branch of the Royal Asiatic Society* (1990) 63, no. 2 (259), 1–15, 12.

[20] Brown, "An Account of a New Genus of Plants named *Rafflesia*," *Transactions of the Linnean Society* (1821), 13: 201–234. Reprinted in *The Miscellaneous Botanical Works of Robert Brown*, vol. I, pp. 367–398. The discovery of the "largest flower in the world" caused a sensation in Europe. Brown's published work in *Transactions* added to his renown. The Dowager Empress of Russia, Maria Feodorovna (1759–1828), was so moved by Brown's account that she sent him a topaz ring as a gift. Brown eventually gave the ring to Lady Raffles in 1856, 2 years before he died. This episode is reported in Bastin's "Sir Stamford Raffles and the Study of Natural History in Penang, Singapore and Indonesia," op. cit., (1990) 12.

[21] There was pressure to name the flower after British naturalists because actually it was first discovered by the French explorer, Louis Auguste Deschamps (1765–1842). Documentation concerning Deschamps's discovery was lost so the British felt it necessary to act quickly.

[22] "List of Plants collected by the Officers, &c., in Captain Ross's voyage, on the Coasts of Baffin's Bay," in *The Miscellaneous Botanical Works of Robert Brown*, Volume I, edited by John J. Bennett (London: published for the Ray Society by R. Hardwicke, 1866–68), pp. 175–178. Reprinted from *A Voyage of Discovery Made under the Orders of the Admiralty, in His Majesty's Ships Isabella and Alexander for the purpose of exploring Baffin's Bay and inquiring into the probability of a North-west Passage* (London, John Murray, 1819) by John Ross, K.S. Captain Royal Navy, Appendix pp. cxli–cxliv.

Fig. 138.—Flower of Rafflesia Arnoldi.

Fig. 11.4 *Flower of Rafflesia arnoldii* by M. Faguet from L. Figuier, *The Vegetable World* (NY: Appleton & Co., 1867), New York Botanical Gardens, The LuEsther T Mertz Library

Fig. 11.5 "The largest flower in the world," *Rafflesia arnoldii* by Franz Bauer, *Transactions of the Linnean Society of London* Vol. 13 (1822), New York Botanical Gardens, The LuEsther T Mertz Library

Scoresby traveled to Spitzbergen and collected different forms of vegetation on that island in northern Norway. Brown published the results of Scoresby's survey as a "Catalogue of plants found in Spitzbergen by Captain Scoresby" (1820). This work listed the plants discovered during Captain Scoresby's voyages in the Arctic, a region that normally would not be expected to contain much vegetation.[23]

The Natural Family, Compositae

Brown expanded his examination of the taxonomy and structure of plants with his work on Compositae (the daisy family, a group of widespread flowering plants), a group he described as "strictly natural," and remarkable "for the great apparent uniformity in the structure of its essential parts of fructification." The Linnean Society (1818) published Brown's paper on Compositae as a monograph as well. In this work, Brown discussed the work of Count Henri Cassini (1781–1832), insisting that although the "near equality of dates, I cannot consider my observations as either wholly or even in any considerable degree anticipated," questioning Cassini's assertion of priority. In spite of his usual reticence, Brown went to considerable length to refute Cassini's claim of priority, demonstrating that he was not to be trifled with when he felt that his integrity was called into question as it had been by Cassini.[24]

Brown went on to describe the fine structure of the flowers of plants in Compositae, indicating "not only the disposition of the five vessels in the tube of the corolla, but their ramification in the laciniae."[25] He also discussed the phenomenon of inflorescence in Compositae. Inflorescence was a term he used to refer to how a group or cluster of flowers were arranged on a stem composed of a main branch, or with a complicated arrangement of branches. He introduced the term "capitulum," which meant the presence of a compact mass of small stalkless flowers (inflorescence), which was found, for example, in the English daisy.[26]

---

[23] Catalogue of Plants found in Spitzbergen by Captain Scoresby, *The Miscellaneous Botanical Works of Robert Brown*, Volume I., pp. 178–179. Reprinted from William Scoresby, *An Account of the Arctic Regions, with a Description of the Northern Whale-fishery*, vol. 2 (Edinburgh: A. Constable, 1820), pp. 75–76.

[24] Brown, *Observations on the natural family of plants called Compositae* (London: Richard and Arthur Taylor, 1817). Extracted from *The Transactions of the Linnean Society of London* 12: 76–142, 76.

[25] Brown, *Observations on the natural family of plants called Compositae*, *The Transactions of the Linnean Society of London* 12: 76–142, 78–79.

[26] Brown, Transactions 12: 76–142, 92–98.

Brown Assesses His Future

In spite of the vast amount of material Banks assigned Brown to work on and catalogue, Brown was growing restless. In 1819 two opportunities emerged, both in his
native Scotland. Daniel Rutherford (1749–1819), Professor of Botany at the
University of Edinburgh, died. Rutherford was a man of considerable intellect and
an accomplished scientist. He was a physician, chemist, and botanist and credited
with the discovery of nitrogen, but he did little to advance the study of Botany at
Edinburgh over the years, and botanical studies there were not in the best of shape
when he passed away. Brown was considered a likely candidate for the position.
Associates from Edinburgh such as Patrick Neill (1776–1851), a horticulturalist and
publisher, several years before (in 1816), suggested that Brown should get his M.D.,
something he would be able to do without much difficulty because of his medical
background. The degree would allow him to gain the necessary credentials for
the post.

Brown hesitated, although with this appointment, he would assume directorship
of the New Botanic Garden in Edinburgh and would be given the opportunity to
make this institution into one of the finest in Britain. His modesty may have been at
play here in his reluctance to accept the position although there was considerable
support for him and perhaps his retiring nature may have been a factor as well. He
never followed up on Neill's suggestion to obtain his M.D.

At the same time, another offer materialized, that of chairman of the Botany
Department at Glasgow University. He turned down both positions however. Banks
had suggested that Brown would inherit his library and collections when he made
Brown his legatee. Perhaps because Brown sensed Banks was not well and was
concerned with the fate of Banks's collections, he felt reluctant to leave him at this
critical time. Eventually William Jackson Hooker—the father of Joseph Dalton
Hooker—took the position at Glasgow University and Robert Graham, Professor of
Botany at Glasgow, received the Professorship at Edinburgh. Banks had in effect
"shackled" Brown "to his possessions forever."[27]

In 1820, Brown's life, the fate of science in Great Britain, and British scientific
exploration were irrevocably altered: Joseph Banks, the longest-serving President
of the Royal Society (over 40 years), passed away. Brown continued his stewardship
as librarian of Banks's collection. He also had the additional responsibility of preserving the magnificent herbarium, library, and drawings in Banks's London home
in Soho Square. The shy and retiring botanist had assumed a preeminent position in
British natural history.

---

[27] "Shackled" is the word David Mabberley used to describe the impact Banks's intervention had
on Brown's career decisions, *Jupiter Botanicus*, p. 216.

Chapter 12
Taking Leave of Sir Joseph Banks

Contents

Banks left his library and collections (particularly his herbarium) to Brown—the product of the work and the investigations of so many dedicated naturalists—with the stipulation that when Brown passed away, all this material should go to the British Museum. In making provision for Brown, Banks allowed for the continuity of the botanical work that had been conducted out of his home in Soho Square, i.e., the use of his library, herbarium, and other materials. Brown received an annuity of £200, providing further financial security, although actually Brown received only half of the annuity for several years after Banks's passing (1820 and 1821). Sir Edward Knatchbull, Lady Banks's nephew and one of the legatees of Banks's estate, initially made it difficult for Brown to carry out Banks's wishes. Because Knatchbull was interested in selling the lease to Banks's house, Brown would be in no position to absorb the cost of renting the entire property.

Brown's colleague and friend, J. E. Smith, saved the day. Smith got the Linnean Society to sublease half of Banks's house while Brown assumed the cost of the other half that housed all the botanical collections. Brown gave up his duties as clerk and housekeeper of the Linnean Society but remained on as its librarian for several more years (until May 23, 1822) when he relinquished his position to an associate, David Don. But Brown retained his salary at the Linnean Society because Smith, still serving as President, had Brown elected as a Fellow of the Society. He could not have become a Fellow while remaining as a paid officer of the Linnean Society. Thus he was relieved of some of his responsibilities without sacrificing the financial stipend that the Society had provided. Furthermore, he fulfilled Banks's wishes stipulated in the latter's will, to continue supervising Banks's vast array of materials. In addition, he had a pleasant and convenient place to live.

Smith and some of Brown's other supporters asked Brown to take the position of Honorary Secretary of the Linnean Society because the Secretary of the time, Alexander Macleay (also spelled McLeay), was resigning to become Colonial

© The Author(s), under exclusive license to Springer Nature Switzerland AG 2021
J. Schwartz, *Robert Brown and Mungo Park*, Memoirs of The New York
Botanical Garden 122, https://doi.org/10.1007/978-3-030-74859-3_12

Secretary of Australia. When Macleay approached Brown with this offer, he did not seem to be too keen to accept. With typical modesty, Brown wrote to Smith, on February 2, 1825, "The Secretary of the Society should unquestionably have the habits of a man of business and be perfectly regular in matters of correspondence." Brown indicated that he did not possess these attributes and doubted that he would ever acquire them. He added that he would be unwilling "to give up so considerable a portion of time as I conceive its duties would require to an office in which after all I might fail in giving general satisfaction or in rendering important service to the Society. If anything could have induced me to decide otherwise, it would certainly have been the gratifying manner in which you have proposed me." Thus he politely and graciously refused the position of Secretary of the Society.[1]

There may have been other reasons for Brown's refusal. There was a faction of members at the Society who felt that Brown was not a skilled administrator and questioned his abilities as a public speaker. Brown may have sensed that there would be some opposition to his appointment as Secretary. In addition, Macleay was not too happy with the arrangement Smith engineered. Macleay was not particularly fond of David Don (1799–1841), the son of one of Brown's earliest mentors from Edinburgh, George Don. David Don took up residence in the part of the house leased by the Linnean Society, one of the job requirements of the position he had taken on.

Macleay and his wife may have been additionally disappointed because their daughter Fanny (Frances Leonora Macleay) was going to go to Australia with them instead of marrying Brown. They assumed that Brown was in love with her and would marry her despite their differences in age. Brown was already 51 and not in the best of health but there is no indication why Brown did not marry her (or anyone else for that matter).[2] Earlier, in 1815, there was a suggestion that Brown had proposed to her but Fanny Macleay's mother, Eliza, was opposed to such a match at the time, apparently much to her later regret.

When Banks left Brown the lease of his home in Soho Square, he assumed that there would be a smooth transition. After Lady Banks's death, Brown was to acquire the rights to the Banks home in Soho Square. The initial problems were luckily solved by Smith's ingenuity, by having the Linnean Society sublease part of Banks's home and Brown could carry on as before. Brown not only received rent for the property the Society leased but he also was allowed to live on the premises at 17 Dean Street, formerly the coach house and stables directly under Banks's library and study.[3]

---

[1] Letter from Robert Brown to James Edward Smith, February 2, 1825, original in the Linnean Society archives, Letter # 14.

[2] There is little information about Brown's personal life but there was speculation regarding his relationship with Fanny Macleay. For example, it was reported in a number of accounts that Brown and Fanny Macleay were engaged in 1815 but there is no information confirming this.

[3] The will stated, "I give and bequeath unto my indefatigable and intelligent librarian, Robert Brown, Esq., an annuity of two hundred pounds .... I also give to the said Robert Brown the use and enjoyment during his life of my library, herbarium, manuscripts, drawings, copperplates engraved, and everything else that is contained in my collection ... and after his decease then I give and bequeath the same to the Trustees for the time being of the British Museum."

Brown now was living next to Banks's marvelous library, drawings, and herbarium. Franz Bauer also received an annuity of £300 a year to continue his work at Kew, and Brown drew on the talents of both Bauer brothers. Brown maintained the productive association he previously had with both Ferdinand and Franz Bauer before Banks's death although Ferdinand left Britain in 1814 to live in Austria. Their insight and artistic talents were immensely helpful to Brown, and Brown called on Ferdinand Bauer's artistic abilities and his insight in natural history. A letter Brown wrote to Franz Bauer on February 24, 1816, several years after his brother's departure from England illustrates Brown's positive relationship with both Bauers. Brown wrote to Franz that he rejoiced in the "progress" his "brother" was making "with his excellent work," referring to Ferdinand Bauer's sketches.[4] In this instance, Brown was soliciting advice on how to best "insert" their sketches in the work he was publishing, most likely his paper on the Compositae (asters and daisies), "Observations on the Natural Family of Plants called Compositae," published in 1817.

Brown as a Clearinghouse in Natural History

After Banks's death in 1820, Brown began to focus on plant structure in his own studies, liberally making use of his microscopic skills. With his position solidified and with firm control of Banks's herbarium and books and drawings, he managed a network of naturalists from Europe and the rest of the world who sent him the results of their experiences in the field. Plant collections continued to pour in from Asia, the Western Hemisphere (particularly South and Central America), the South Pacific, and the Arctic regions. In 1823, Brown published *A List of Plants collected in Melville Island in the year 1820; by the officers of the Voyage of Discovery under the orders of Captain Parry, with characters and descriptions of the New Genera and Species.*

Captain William E. Parry (1790–1855) traveled to Melville Island in the Canadian Arctic in 1819–1820, searching for the North West Passage. A supplement to Captain Parry's Voyage was published, and in this work Brown listed the plants collected during the expedition. Brown's commentary on the list was written entirely in Latin except for his introductory remarks. Brown acknowledged the use of Captain Edward Sabine's (1788–1883) herbarium, and explained that he took so much time in publishing due to the state of his own health. He indicated that the plants of Melville Island were found in the ratio of five dicotyledonous plants to two monocotyledonous plants, an unusually low ratio of dicots to monocots. This may

---

[4] Letter from Brown to Francis Bauer, February 24, 1816, Botany Library, Natural History Museum.

have been because there was an unexpectedly large number of grasses, which are monocots, "nearly double of that in any other part of the world."[5]

Walter Oudney (1790–1824) was a Scottish physician and African explorer. He received his degree from Edinburgh University (1817).[6] In early 1822, Oudney departed from Tripoli with explorers Dixon Denham (1786–1828) and Captain Hugh Clapperton (1788–1827). They reached Bornu in February 1823, and thus they became the first Europeans to accomplish a north-south crossing of the Sahara Desert, a journey that took a year's time. Stricken by illness—probably dysentery—Oudney died in January 1824 in the village of Murmur, located near the town of Katagum. There are certain parallels with Mungo Park's exploits although Oudney's travels in northern Africa were of shorter duration but perhaps more productive from the perspective of natural history.

On his journey, Oudney collected regional plants, a total of approximately 300 species. 100 species were collected from the area near Tripoli, 50 from the central region of Libya to the south, 32 from Fezzan, another 33 on route to Bornu, 77 in Bornu, and finally 16 in Hausa. Some of this material was lost when it was sent to Brown. Brown discovered that out of all the material he received, only a few of the species were found to be new or previously not recognized—fewer than 20. Brown found no new genera among the plants sent to him from the expedition. Due to the paucity of new species in the collection, Brown decided not to publish a list of them but instead published the observations Oudney and others had made on the structure and affinities of the more remarkable or unusual plants they had encountered. Brown named a monocotyledonous genus belonging to the Brassicaceae—previously unnamed—*Oudneya*, in Oudney's honor.

In 1826, the two-volume *Narrative of Travels and Discoveries in Northern and Central Africa in the years 1822, 1823, and 1824* was published describing the African exploits of Oudney, Denham, and Clapperton. Brown wrote an appendix to this work, *Observations of the Structure and Affinities of the More Remarkable Plants collected by the Late Walter Oudney, M.D. and Major Denham and Captain Clapperton in the years 1822, 1823, and 1824, during their Expedition to Explore Central Africa.*[6] Brown found that reviewing material sent to him by other naturalists and getting the results published were becoming tedious. He felt the need to devote more time and energy to his own studies.

---

[5] Captain Edward Sabine (1788–1883) compiled lists of mammals, birds, fish, and marine invertebrate animals. Sabine was an Anglo-Irish explorer, a soldier who served on voyages of exploration with Captain John Ross and Captain Parry. He was also an ornithologist, geophysicist, and astronomer, and eventually became the 30th President of the Royal Society. Other naturalists assembled lists for land invertebrates, shells, and rock specimens. *A supplement to the Appendix of Captain Parry's Voyage for the Discovery of the North-West Passage in the Years 1819–20* (London: John Murray, 1824), pp. cclxi–cccx. Brown's introductory remarks provide a good overview of the botanical observations, pp. cclxi–cclxii.

[6] Dixon Denham, Hugh Clapperton, Walter Oudney, Abraham V. Salamé, *Narrative of Travels and Discoveries in Northern and Central Africa in the years 1822, 1823, and 1824* (London, John Murray: 1826), two volumes. Brown's supplement was published as *Observations of the Structure and Affinities of the More Remarkable Plants collected by the Late Walter Oudney, M.D. and Major Denham and Captain Clapperton in the year 1822, 1823, and 1824, during their Expedition to Explore Central Africa*, in *The Miscellaneous Botanical Works of Robert Brown* (London: Thomas Davison, 1826). Before his death, Oudney had been appointed consul by the British Government for promotion of trade with the Kingdom of Bornu in sub-Saharan Africa.

Brown began to refine his own work, not only drawing on the material sent to the Royal Society when it was useful for his own investigations. He spent considerable time on the histology of cycads and conifers, acutely examining the tissues of these plants. He explored the relationship between the fine structure of cycads and conifers, specifically their reproductive structures. He had begun this work while Banks was alive (before 1820) but he had put this work aside, not publishing his observations in detail until later on, i.e., the 1820s and 1830s. He needed time to reflect on his anatomical observations, deferring his work on the fine structure of these seed plants. Eventually, the careful observations he made allowed him to grasp differences between primitive cryptogams (mosses and liverworts) and the more advanced seed plants, gymnosperms and angiosperms. He observed the advantage seed plants had in reproduction. They did not require moisture to reproduce. His knowledge concerning the alternation of generations among different groups provided Brown insight into how seed plants were better adapted to a wider range of environmental conditions but obviously he did not have the level of sophistication later botanists had. However, his morphological observations were very useful in supporting his taxonomic work.

While Brown continued his investigations, his colleague, J. E. Smith, was busy on his study of English flora, confiding to the Bishop of Carlisle that he was working on grasses, specifically *Triandria Monogynia* (saffron). He remarked, "Brown is very good in these," again illustrating the high regard Smith had of Brown. The relationship between Smith and Brown remained a very positive one and their collaboration yielded productive results in descriptive or observational botany. It also showed how much Smith (and others) relied on Brown's expertise in a whole range of questions that arose.[7]

In his 1825 paper presented before the Linnean Society, Brown broke new ground concerning his work on cycads and conifers. He attached his paper "Character and Description of Kingia; a new genus of plants found on the southwest coast of New Holland with observations on the structure of its unimpregnated ovulum, and on the female flower of Cycadeae and Coniferae" to King's published *Narrative of a Survey of the Intertropical and Western Coasts of Australia*. Philip Parker King (1791–1856) had made multiple trips surveying the coasts of Australia, particularly the western coasts, gathering information on the flora and fauna, the topography, the minerals and timber resources, the climate, and the ethnography of these regions.

Brown commented, "To this new genus I have given the name of my friend Captain King, who, during his important surveys of the Coasts of New Holland, formed valuable collections in several departments of Natural History, and on all occasions gave every assistance in his power, to Mr. Cunningham [Alan Cunningham

---

[7] Letter from Smith to the Bishop of Carlisle, March 3, 1822, James Edward Smith, *Memoir and Correspondence of the Late Sir James Edward Smith*, ed. by Lady Pleasance Reeve Smith (London: Longman, Rees, Orne Brown, Green and Longman, 1832), vol. I, p. 602.

(1791–1839)], the indefatigable botanist who accompanied him. The name [Kingia] is also intended as a mark of respect to the memory of the late Captain Philip Gidley King, who, as Governor of New South Wales, materially forwarded the objects of Captain Flinders' voyage, and to whose friendship Mr. Ferdinand Bauer and myself were indebted for important assistance in our pursuits while we remained in the colony." This was quite a generous encomium to the former governor, in view of the friction that had sometimes transpired between Brown and King 20 years before, most particularly over Brown's hesitation to travel back to England aboard the *Investigator*, and King's insistence that he do so.[8]

Brown advanced the idea that the cycads and conifers contained ovules that were not "imbedded" in their ovaries but were visible to even casual observers, an observation he had made previously. His addendum to King's work provided Brown with an ideal forum to more fully present his ideas. This led to the discovery, more fully enumerated by later botanists, that these groups of plants possessed seeds (ovules) that were exposed ("naked") unlike the covered seeds (ovules) found in flowering plants that are not readily visible to the unaided eye. He thereby grasped the essential difference between gymnosperms (cycads and conifers) and angiosperms (flowering plants). Brown wrote, "the greater simplicity in Cycadeae, and in the principal part of Coniferae of the supposed ovulum which consists of a nucleus and one coat only, compared with the organ as generally existing when enclosed in an ovarium."[9] Up to this time, botanists had classified conifers and cycads as flowering plants, instead of using the modern terminology. Although conifers and cycads are plants bearing seeds like flowering plants, they are not flowering plants, a distinction not clear at the time or properly elucidated. Brown eventually placed cycads and conifers in a separate group apart from "phaenogamous" plants, i.e., true flowering plants, the angiosperms. In spite of this insight, when he published "Kingia, a new genus of plants," he continued the practice of referring to cycads and conifers as flowering plants. Brown again was demonstrating his usual caution, hesitating to break completely with traditional methods of putting the plants he had observed into their appropriate groups until obtaining further evidence.

---

[8] From page 3 of the privately printed copy of *Character and Description of Kingia*. Philip Parker King was the son of Captain Philip Gidley King, who was Governor of Australia during the time Brown conducted his explorations there.

[9] Page 24 of privately printed copy of paper, page 454, of *Miscellaneous Botanical Works*, ed. by John Joseph Bennett, "Character of Kingia, a New Genus of Plants found on the South-west Coast of New Holland: with Observation on the structure of its Unimpregnated Ovulum; and on the Female Flower of Cycadeae and Coniferae," first read before the Linnean Society on November 1 and 15, 1825. A limited number of copies were published in 1827 and then were republished later on in John J. Bennett's compendium, *The Miscellaneous Botanical Works of Robert Brown*.

The Politics of the Royal Society

Brown began to be involved in the politics of the Royal Society in the decade following Banks's death (1820s). John Frederick William Herschel (1792–1871), mathematician, astronomer, chemist, a pioneer in photography, and an inventor as well as a botanist, led efforts to wrest control of the Royal Society from scientists belonging to the aristocracy—either aristocratic by birth or who had aligned themselves with the aristocratic establishment. Sir Humphrey Davy (1778–1829) had succeeded Banks as President and continued to represent the interests of the "old guard." As heir to Banks and his scientific estate, one might assume that Brown would side with Davy and the scientific establishment but this notion is counterintuitive. Brown supported the "young Turks" in their efforts to inject fresh blood into the organization as did chemist William Hyde Wollaston (1766–1828). Wollaston was a colleague who worked with Davy but nevertheless found that he was unable to continue his support. The insurgents failed in their initial efforts to establish new leadership although they lost the first election by only a slender margin.

Eventually they succeeded in their attempt to breathe new life into the Society. Davy resigned in 1827 when he lost the support of Michael Faraday (1791–1867) who was an admirer of Herschel. In hindsight, it becomes more apparent that although Brown was loyal to Banks and his memory, he was very much his own man and expressed his views in a quiet but firm manner. It also illustrates that when he believed in the value of a particular cause, he did not equivocate, and it was very much comparable to how he maintained his position when he abandoned the widely accepted Linnean system of classification in favor of the "foreign" system.[10]

Brown's adoption of a more natural system of classification did not affect his standing with his colleagues and he gained even more respect for the courage of his convictions. When J. E. Smith wrote to Nathaniel Wallich (1785–1854), a Danish-born physician and botanist and expert on the plants of India, he commented on the large number of plants Wallich had sent him in May of 1819. Smith declared in his letter of March 6, 1820, to Wallich that he defended "the Linnaean fortress as stoutly as I can; but I am thankful for, and concur in, all sound correction and improvement. Mr. Brown and I are sworn friends, though he uses the natural arrangement. He is one of the most amiable, acute, and worthy of men."[11]

In 1826, Brown decided not to wait to transfer Banks's library and other material to the British Museum although Banks had stipulated that the transfer should be made after Brown's death. Brown felt that this marvelous collection belonged to the nation, and so he struck a bargain with the trustees of the Museum. All the material

---

[10] For a full account of this episode, see David Philip Miller's "Between Hostile Camps: Sir Humphry Davy's Presidency of the Royal Society of London, 1820–1827," *The British Journal for the History of Science* (1983) 16 (1): 1–47.

[11] Letter from J. E. Smith to Nathaniel Wallich, March 6, 1820, James Edward Smith, *Memoir and Correspondence of the Late Sir James Edward Smith*, ed. by Lady Pleasance Reeve Smith (London: Longman, Rees, Orne Brown, Green and Longman, 1832), p. 261.

would be transferred to the British Museum, thus forming an independent department in the Museum. He would take charge of it as under-librarian, in effect, "Keeper of the Botanical Department of the British Museum." He spent a good part of 1826–1827 moving Banks's library, herbarium, and other collections to the British Museum from Soho Square, thus helping to establish Britain's first nationally owned botanical collection accessible to the public. Brown remained in charge of this collection at the Museum until his death in 1858, and he continued to reside at Banks's former home in Soho Square.[12]

Brown's hand had been essentially forced in this matter by a series of events not of his making. During this period, the Museum and England's great universities, particularly Oxford and Cambridge, were in serious decline. To address this crisis, numerous political figures and members of the intellectual elite proposed that a new university should be created in London, a university where there would be no requirement for members to belong to the Church of England and would be free of the medieval trappings characteristic of Oxford and Cambridge.

Brown was offered the Professorship of Botany there. But there was a catch. Under the terms of Banks's will, Brown would have to surrender his annuity and Brown would have to cede his control of Banks's library and collections. The Museum would gain this valuable resource with no financial obligations or any other commitment to Brown. Brown was unwilling to agree to these terms. After a complex series of negotiations, Brown decided not to take the Professorship at London University but agreed to the transfer of Banks's botanical collections to the Museum where they would remain under his control. Brown not only made a decision that was personally advantageous, but he was also abiding by the Banks's wishes that were expressed in his will. The British Museum (Natural History) gained as well. Having a botanical collection in their premises that was to be fully accessible to the British public would invigorate the Museum's Department of Botany.[13]

---

[12] Brown transferred almost everything except his own private herbarium that he had established while conducting his own researches.

[13] In reality, Brown maintained tight control over the collection. Phyllis I. Edwards indicates, "Unlike Banks, Brown did not allow free access to *his* collection; this was still true even when it came into the possession of his colleague and successor at the British Museum, J.J. Bennett (1801–76)," *Journal of the Society for the Bibliography of Natural History* 7 (1976): 385–407, 405.

Chapter 13
Pollen Grains of *Clarkia pulchella*

Contents

Although Brown continued to work on his taxonomic studies (which helped influence the growth of the natural system of classification in England), he picked up the threads of research that he had begun earlier, studying the minute structure of the vegetative and reproductive organs of plants. It was while he conducted this work that he made an important discovery concerning the behavior or action of minute particles, a phenomenon that piqued the interest of scientists from a variety of backgrounds. This was the discovery of the phenomenon of "Brownian motion," an achievement that epitomized his journey from a descriptive naturalist to an investigative scientist, one who was less reluctant to speculate on the vast amount of information he had accumulated over many years of careful study.

Brown accomplished further breakthroughs during this period (from the late 1820s to the 1850s). His recognition of the "naked" (uncovered) condition of the ovules in the cycads and conifers eventually led to the taxonomic division of seed plants into two distinct groups, the gymnosperms, conifers and cycads, and flowering plants, the angiosperms. He also conducted significant work in cytology, developing a deeper understanding of the composition of the cell. He identified the phenomenon of protoplasmic streaming. Perhaps most important was Brown's observation of the nucleus in certain plant cells, an observation made by several other pioneers in cytology although Brown did not grasp the full significance of his discovery at the time. He would spend the majority of his remaining years on such matters.

© The Author(s), under exclusive license to Springer Nature Switzerland AG 2021
J. Schwartz, *Robert Brown and Mungo Park*, Memoirs of The New York
Botanical Garden 122, https://doi.org/10.1007/978-3-030-74859-3_13

Brownian Movement and Brown's Microscopy

While Brown was conducting microscopic examination of plant structure, he made a series of observations on the nature and behavior of microscopic particles suspended in fluids. He determined that their movements did not seem to possess a "specifically organic character," and believed that it was worth further investigation.[1] Brown was not the first to observe the motion of microscopic particles—in this case, pollen grains of the flowering plant, *Clarkia pulchella* (commonly called elkhorns or ragged robin, pink fairy, clarkia or deerhorn)—in fluids such as water. But more significantly, he had demonstrated that any kind of matter, whether organic or inorganic and reduced to smaller particles, would exhibit random movement when suspended in a medium like water. This noteworthy observation proved to be relevant to biologists and physical scientists as well. Brown was drawn to this subject because he became curious about the movement of pollen particles in fertilizing the ovum and wanted to learn more about this phenomenon.[2]

Brown's paper, *A Brief Account of Microscopical Observations Made in the Months of June, July and August, 1827, on the Particles Contained in the Pollen of Plants; and on the General Existence of Active Molecules in Organic and Inorganic Bodies*, was privately printed. It was subsequently published in the *Edinburgh Journal of Science* in 1828 as well as reprinted elsewhere although initially Brown only wanted it to be published privately because of its speculative nature.[3] In his investigations, Brown used his trusted simple microscope that was capable of magnifications up to 300 diameters, even though the use of compound microscopes began to be more common at the time. Although greater resources were now at his disposal with the promise of achieving greater magnification, Brown remained more comfortable using his simple microscope (Figs. 13.1 and 13.2).

The pollen Brown suspended and observed was from 1/4000th to 1/5000th of an inch in diameter. He noted, "While examining the form of these particles immersed in water, I observed many of them very evidently in motion; their motion consisting not only of a change of place in the fluid, manifested by alternation in their relative positions, but also not infrequently by a changes of form of the particle itself …

---

[1] See Stephen G. Brush, "A History of Random Process, I. Brownian Movement from Brown to Perrin," *Archive for History of Exact Sciences* 5 (1968): 1–36, pages 1 to 5 are relevant to this discussion. Brush's account is from a different perspective, a historical examination by a physicist and historian of the physical sciences.

[2] William Clark of the Lewis and Clark expedition in the United States found this plant in the Bitterroot Mountains in Idaho, and he turned over his botanical specimens to the German-American botanist, Frederick Traugott Pursh (1774–1820), who gave the plant its scientific name, *Clarkia pulchella*, in honor of the explorer Clark; the species name *pulchella* meant beautiful.

[3] Among the journals where it appeared were *Edinburgh New Philosophical Journal* (1828) 5: 358–371 and *Annales des Sciences Naturelle* (Paris) (1828) 14: 341–362.

Fig. 13.1 Robert Brown's microscope, restored by Professor Brian J. Ford, honorary surveyor of scientific instruments at the Linnean Society of London, Permission of the Linnean Society of London

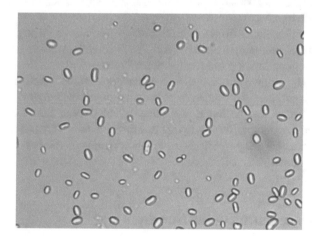

Fig. 13.2 Pollen grains of *Clarkia pulchella*, © 2010 by P. Pearle, K. Bart, D. Bilderback, B. Collett, D. Newman, S. Samuels

after frequently repeated observation, that they arose neither from currents in the fluid, not from its gradual evaporation, but belonged to the particle itself."[4]

---

[4] *A Brief Account of Microscopical Observations Made in the Months of June, July and August, 1827, on the Particles Contained in the Pollen of Plants; and on the General Existence of Active Molecules in Organic and Inorganic Bodies.* Copy is in the LuEsther T. Mertz Library of the New York Botanical Gardens.

Brown did not understand the actual cause of the random motion of the particles. He rejected the notion that this motion was caused by the mutual attraction or repulsion of the suspended particles and the water molecules. Brown noted the same motion with nonliving glass and rock particles that he had observed with pollen. He erroneously maintained that the movement "belonged to the particle itself," i.e., its movement was autonomous. Brown's limitations in his grasp of the physical sciences probably handicapped him in determining why suspended particles behaved in the manner he observed.[5] However, Brown deserves credit for ascertaining that matter, inorganic or organic, when broken down into finer particles but still visible microscopically, would exhibit discernible random motion when suspended in water or other fluids. His observation was considered a significant contribution at the time and resulted in having the phenomenon named after him.

Brown's elucidation of the random movement of suspended particles is an example of how he carefully analyzed problems that intrigued him. Charles Darwin understood the serious attention Brown gave to his scientific investigations. Darwin recalled his frequent meetings with Brown in the 1830s when he was a young man, calling on him before he departed on his voyage of the *Beagle* to hone his microscopic skills. Darwin cited Brown's extraordinary knowledge in his *Autobiography* years later (1875). He worried that although Brown "poured out his knowledge to me in the most unreserved manner," he was too cautious, "owing to his excessive fear of ever making a mistake." As a result, Darwin feared much of Brown's knowledge died with him.[6]

When Darwin went to Brown to be tutored in microscopy, Brown showed him the protoplasmic streaming he had observed in plant cells. Darwin later built upon Brown's work in pollination and fertilization in flowering plants. Brown's discoveries, particularly through the use of his microscopic skills, made an impression among prominent men and women even in the nonscientific community. A notable reference to Brown can be found in Chapter 17 of George Eliot's *Middlemarch*. There, the young doctor, Tertius Lydgate, offered a trade, "And I will throw in Robert Brown's new thing—'Microscopic Observations on the Pollen of Plants'— if you don't happen to have it already."[7]

Brown's Observation of the Nucleus

Brown continued his microscopic investigations after his work on the random movement of particles. In 1831, he privately published a pamphlet, "Observations on the Organs and Mode of Fecundation in Orchidae and Asclepiadeae," reprinted

---

[5] In 1905, Albert Einstein suggested that Brownian motion was the result of the particles colliding with (water) molecules. Brown's discovery provided the first evidence suggesting the existence of atoms.

[6] *Charles Darwin: His Life told in an autobiographical Chapter, and in a selected series of his published letters*, edited by his son, Francis Darwin (London: John Murray, 1892), p. 46.

[7] George Eliot, *Middlemarch, A Study of Provincial Life* (Boston and New York: Houghton Mifflin Company, 1908), p. 251.

and published in the *Transactions of the Linnean Society of London*.[8] In this work, Brown discussed the behavior of orchid pollen when placed on the stigma of the flower's pistil. He observed that they produced pollen tubes that grew down into the ovary. His primary object in conducting this line of research almost pales in comparison with the important finding that resulted from his labors, and which had great significance as far as cell theory and rudimentary developments in the field of cytology were concerned, i.e., his discovery of nuclei in cells of the leaves of the orchids. On November 1, 1831, he presented "The discovery of the nucleus of the vegetable cell" before the Linnean Society, reporting his important microscopic observations.[9] He described the division of the pollen mother cells of *Tradescantia* in the formation of tetrads and the maturation of pollen grains. He verified the earlier findings of Giovanni Battista Amici (1786–1863) and Adolphe Théodore Brongniart (1801–1876) who described the formation of the pollen tube. Brown reported that the tubes grow toward the micropyles of the ovules. He was uncertain whether all the tubes grew out of the pollen grains or were indirectly formed from them. His uncertainty about the origin of the pollen tubes was that his explanation was "not altogether satisfactory," a remark exhibiting a combination of characteristic modesty and candidness. Referring to the leaves of orchids cells, Brown observed that in "each cell of the epidermis ... [there was] a single circular areola, generally more opake than the membrane of the cell ... and only one areola belongs to each cell .... This areola, or nucleus of the cell ... is not confined to the epidermis ... but in many cases in the parenchyma or internal cells of the tissue."[10]

Perhaps, the most noteworthy aspect of this paper was Brown's observation that certain plant cells contain a dark-staining body, the nucleus. He was not the first investigator to note its presence. In 1781, Felice Fontana (1730–1805) reported nuclei in the skin cells of eels, and in 1700, Antonie van Leeuwenhoek reported "retractile" bodies in the center of white blood corpuscles of salmon. Brown might not even have been the first to identify nuclei in plant cells but he was the first to give a full description of this body. He made a series of observations of the nuclei of cells in many parts of the same plants. He focused on plants belonging to certain families, particularly among monocotyledons, but occasionally selected samples of plants from dicotyledonous families. Typically, Brown was cautious by not stating that the nucleus (or "areola" as he sometimes called it) was found in all cells.

However, Brown avoided much discussion of when and where the nucleus could not be observed. Brown attributed its absence to the shortcomings of the observer, suggesting that it could be found if it was sought after more skillfully, thereby acknowledging his own shortcomings. With his observations of nuclei in the cells of

---

[8] Robert Brown, *Transactions of the Linnean Society of London* 16 (1833): 685–745.

[9] Read before the Linnean Society on November 1 and 15, 1831.

[10] Brown, "Observations on the Organs and Mode of Fecundation in Orchideae and Asclepiadaceae," *Transactions of the Linnean Society of London* 16 (1833): 685–745, 707. See Vasco Ronchi's "Amici, Giovanni Battista," in *Dictionary of Scientific Biography*, vol. 1 (New York: Charles Scribner's Sons, 1970), pp. 135–137, and M.J.S. Rudwick's "Brongniart, Adolphe Théodore," *Complete Dictionary of Scientific Biography*, vol. 4 (2008).

ovules, pollen mother cells, and tetrads, Brown was on the threshold of accounting for the nucleus's role in cell reproduction. Perhaps his natural caution did not allow him to make this final leap and fully explain this phenomenon. Nevertheless, Brown's discovery represents an important advance in cytology.

Brown disclosed that the nucleus was not confined to the Orchideae but was "equally manifest in many other Monocotyledonous families." He did add that he found it "in the epidermis of Dicotyledonous plants," but reported, "I have found it hitherto in very few cases in the epidermis of Dicotyledonous plants."[11] Other botanists also reported seeing nuclei in specific plant tissues but Brown was the first to note its presence in a wide variety of cells, and may have been the first scientist to name this "opake" (opaque) body. Obviously among the pioneers in the development of the cell theory, Brown should be included.

In the 1830s, Brown began to receive recognition for his accomplishments: he was elected as one of eight foreign associates of the Academy of Sciences of the Institute of France. He already had received the honorary degree of D.C.L. (Doctor of Civil Law) from Oxford University in 1832 along with John Dalton, Michael Faraday, and Sir David Brewster. In a similar vein, in 1839, the Council of the Royal Society awarded him the Copley Medal, the highest honor they gave, in recognition of his work in plant fertilization. There were other honors of that nature for Brown. Brown was ready to move from the result of work he had accomplished in nineteenth-century descriptive botany, where he already achieved status as a towering figure, to doing more speculative studies. But often he was required to put that aside because he was pressed to catalogue the material sent to him by other naturalists, especially after Banks's death.

Work on Thomas Horsfield's Observations

The last important taxonomic work undertaken from the material other naturalists sent to Brown was *Plantae Javanicae Rariores*, an effort that was published in four parts, in the years 1838, 1840, 1844, and 1852. For this project, Brown had the assistance of John J. Bennett, his coauthor and his eventual successor at the Natural History Museum. The study was based on the collections made in Java by Thomas Horsfield (1773–1859) between 1802 and 1818. Horsfield's work was a prodigious undertaking, surveying 2196 species.

Horsfield, an American doctor—from Bethlehem, Pennsylvania—first worked for the Dutch East India Company in Java in 1801 in order to study the medicinal plants used by the natives in that colony. This distant land first attracted Horsfield's attention when he served on the Philadelphia-based merchant ship, *China*, on a visit to Batavia (Java). He was quite taken by the beauty of Java's scenery and the

---

[11] "Observations of the Organs and Mode of Fecundation in Orchideae and Asclepiadaceae," *Transactions of the Linnean Society of London* 16 (1833): 685–745, 710–712; also in the privately printed manuscript, "Observations" (1831), 19–21.

grandeur and abundance of its vegetation, so much so that he returned frequently during the next 16 years, studying the flora and fauna as well as the geology of Java and the adjoining areas like Sumatra. He secured a position as surgeon for the Dutch colonial government, and then resumed working for the East India Company. During that period of time, his work in natural history attracted the attention of the Colonial Governor, Sir Stamford Raffles. Raffles wrote to Banks a letter of introduction on Horsfield's behalf in 1818, explaining that Horsfield's collections were immense and that he was coming to England.[12]

In 1819, Horsfield moved to England and became Assistant Librarian for the East India Company. Horsfield collected many plants between 1802 and 1818, amassing considerable fauna, antiquities, and minerals as well as plants. He sent his botanical specimens to Banks with the hope that Brown would prepare a detailed enumeration of them fairly soon afterward. However, Horsfield was initially more successful in publishing his observations on Javanese animals, *Zoological researches in Java and the neighboring islands* (1824), and this work established his reputation as a naturalist.[13]

The success of Horsfield's publication of Javanese animals served as a prelude to the large botanical work on the Javanese plants he planned, *Plantae Javanicae Rariores*. Horsfield chose Brown to examine and classify his botanical collection because he knew how successful Brown had been in taking naturalists' collections before, and then turning them into well-polished and important publications. However, in this instance, Brown took some time assisting Horsfield in naming and classifying the new species of plants Horsfield observed and collected in Java and had shipped to him. Brown's apparent lack of enthusiasm in working on Horsfield's collection stands in contrast with the earlier previous work he did with other material sent to him, and it has been noted by commentators.[14]

---

[12] Letter from Raffles to Banks, August 14, 1818, *D.T.C.* 20: 105–107. Also in *The Indian and Pacific Correspondence of Banks,* vol. 8. The letter was written shortly after Joseph Arnold's discovery of *Rafflesia arnoldii* and Arnold's death.

[13] See Robert McCracken Peck's note, "Discovered in Philadelphia: a third set of Thomas Horsfield's nature prints of plants from Java," in *Archives of Natural History* (2014), 44(1): 168–170.

[14] Botanist and historian, William T. Stearn criticized Brown's lack of interest in tackling Horsfield's collection, indicating that while "Horsfield … hoped for a complete enumeration of his material classified and named by Brown," it was delayed by "Brown's other activities and his eleven-week holidays." This may be an unduly harsh criticism because *Plantae rariores Javanicae*, when it was published, represented a considerable contribution, although Stearn commented that it consisted of "only" 258 pages and 50 plates, and Brown contributed written commentary on 30 of Horsfield's 2196 species. "Robert Brown," in *Dictionary of Scientific Biographies*, volume 2, pages 516–522 (1970), pp. 520–521. However, Stearn was effusive in his praise of Brown's skills in his introduction to a modern edition of *Prodromus Florae Novae Hollandiae*. He wrote: "The magnitude of Robert Brown's contribution of the Australian flora as collector and investigator becomes evident when set against what was known about Australian plants to the end of 1805, the year of his return to England from Australia," Stearn, "An Introduction to Robert Brown's *Prodromus Florae Novae Hollandiae*," Weinheim: Engelmann, 1960.

When *Plantae Javanicae Rariores* was eventually published, it marked a considerable effort; it contained 258 pages and 50 plates of illustrations. Brown only contributed written commentary on 30 of Horsfield's 2196 species included there. Bennett and others worked on the remainder under Brown's direction but the project took considerable time, over the course of many years, 1838–1852.

Brown clearly wished to focus on "those subjects" he was most interested in and now felt that he had the freedom to be more selective in devoting his energy and time to matters he considered more important to him. A contemporary of Brown, gardener and botanist, John Lindley (1799–1865), criticized Brown for his "habits of procrastination," in connection with Horsfield's work.[15] Lindley may have another motivation for his harsh criticism of Brown. He and Brown differed sharply on taxonomic questions, with Lindley advocating that the physiology of plants should be a criterion in classification in addition to plant morphology and histology, while Brown indicated that it would lead to ambiguity or confusion. Lindley, years earlier as a young man (1826–1827), took the Professorship of Botany at the newly founded University of London, a position that Brown had been offered and refused.

Plantae Javanicae Rariores is now regarded as a classic of Indonesian botany, and Horsfield's specimens are preserved in the Department of Botany of the Natural History Museum. It was perhaps unfair to judge Brown too harshly in light of the charges of procrastination that was leveled against him at the time, a practice continued by some contemporary botanists. Expecting Brown to give priority to Horsfield's specimens at the same time he was busily engaged with other problems after Banks's death, may be too rash a judgment. Brown was busily working on the collections of other naturalists at the time as well. He may have felt that it was an imposition to have to deal with still another shipment of plant collections that represented the work of others while he wanted to conduct his own investigations and not take on the arduous task of completing the work of others. Further, Horsfield was quite positive when it came to Brown's efforts, indicating in a postscript to his work, "Mr. Brown cheerfully devoted particular care and attention, and his analysis of vegetable structure will be duly appreciated by Botanists. After the completion of the drawings and of the illustrative details, the subjects were put into the hands of the engraver, and Mr. Brown commenced the preparation of the text." He added "his great obligations to Mr. Brown. The examination and arrangement of my Herbarium, the laborious duties connected with the superintendence of the figures contained in this work, the preparation of the illustrative details, and the time devoted to the description of the subjects, are by no means the only marks of friendship which I have received from that distinguished Botanist."[16]

---

[15] *Gardener's Chronicle*, 1852 (26 June): 406–407.

[16] See *Plantae rariores Javanicae*, Thomas Horsfield, M.D., Robert Brown and John J. Bennett ed. *The Miscellaneous Botanical Works of Robert Brown*, pp. 560–561.

Brown's Final Years

Brown's focus during the last few decades of his life was in paleobotany, earlier marked by a fossil-hunting trip he took to the Burgundy region of France in 1836. He also received fossils from Darwin, including wood that Darwin collected from Valparaíso, Chile, while on the *Beagle*. His natural caution may have limited him in speculating on the significance of the fossils he was examining. He did not tie this work to the possibility that these specimens were evidence of organic change. Brown continued to publish work describing in minute detail the fine structure of plant fossils, illustrated by his 1847 paper on the strobilus of the cone of an extinct fossil plant, a treelike organism related to the Lycopsids or club mosses belonging to the genus *Lepidodendron*. It was named *Lepidodendron brownii* in his honor, and commonly referred to as "Brown's cone."[17] Members of present-day Lycopsida are relatively small-sized plants, surviving representatives of these early vascular plants that were quite large.

Brown's devotion to paleobotany in his later years may indicate that he toyed with the idea of accepting some form of gradual evolutionary change although he did not publish anything on the subject. His rudimentary thoughts on speciation would be a logical result of the taxonomic studies he had accomplished over the years. Brown's opportunity to travel to exotic lands far from the carefully landscaped England in addition to the rugged landscape of his native Scotland may have been influential in allowing him to accept ideas that were inimical to the idea that species were immutable. It seems likely that he read or was aware of more popular works such as Robert Chambers's anonymously published *Vestiges of the Natural History of Creation* as well as other serious discussions about the possibility of organic change. Brown generally did not comment on popular works such as *Vestiges* and on what he read in the newspapers and other periodicals of the day.

However, an exception to this occurred on February 17, 1845, in an exchange Brown had with Sir Charles J. F. Bunbury (1809–1886), at an evening party held at Bunbury's home. Bunbury, the eighth Baronet, was a naturalist with an active interest in geology and paleobotany as well. He was a good friend of Charles Lyell (and related to Lyell by marriage), and had accompanied Lyell on a trip to Madeira. He also collected plants in South America in 1833 and South Africa in 1838 and was keenly interested in the possibility of the transmutation of species. Bunbury reported in his journal on the same day of his party that he "worked for three hours at fossil plants in the Museum of the Geological Society." Bunbury asked Brown about the transmutation of oats into rye and "one sort of corn into another," alluding to a "much discussed" note published in *Gardener's Chronicle* at the time. Bunbury reported that Brown "would not express decidedly any opinion of the subject (according to his manner), but smiled sarcastically, and remarked that such a transmutation might be *very convenient*. He said the author of *Vestiges of Creation* has

---

[17] See *The Miscellaneous Botanical Works of Robert Brown*, J. J. Bennett, ed., 2 vols. (London, 1866–1867).

availed himself of this supposed fact, imagining it to be favourable to his theory," i.e., gradual change. Brown went on to explain that "it is in reality adverse; for his theory is that of the gradual development of forms, whereas the fact in question would be an instance of an alteration *per saltum* from one form to another considerably different."[18]

This was all Brown ever said about the popular work, *Vestiges*, at least publicly. Interestingly, Brown believed that the author of the *Vestiges* was advocating gradual change but to many observers at the time, the change seemed rapid, anything but gradual. Here again, one senses that Brown might have offered his opinion regarding the possibility of change in the structure and form of species, but he was reticent in expressing himself on the subject. His remarks to Bunbury indicated that he may have been more inclined to support some type of gradual change but Brown did not allow himself to say anything more on the subject.

Brown maintained his association with such old colleagues and friends as Alexander Macleay. In 1840, after not hearing from Macleay for some time ("3 & 4 years since I receiv'd your last letter"), Brown wrote to him that "I begin to dispair [sic] of hearing from you again … nevertheless I continue to write now & then as opportunities offer. My present object is to recommend to your kind attention Mr John Lyall a cousin of my own & of Mr Charles Lyall's [no relation to the famous geologist] who is on the point of embarking for Sydney as an agent in connection with the House of Lyall's Brothers here."[19]

Brown spent much of his time after Banks's death in supervising the collections at the British Museum (Natural History) and trying to acquire new botanical material. For many reasons, primarily his natural diffidence and modesty, he initially rejected an offer to be nominated as President of the Royal Society. It was only after many entreaties that he finally agreed to become President of the Linnean Society in 1849, serving in that capacity until 1853. His position as Keeper of the Botany Department of the British Museum was handicapped by a lack of funds needed to purchase new collections offered to the Museum. For example, the Russian herbarium of John D. Prescott—a merchant in the city of St. Petersburg, Russia, where he died in 1837—containing approximately 25,000 specimens became available in the early 1830s. The trustees of the Museum required only an appropriation of £1000 to

---

[18] Charles James Fox Bunbury, *The Life of Sir Charles J.F. Bunbury, bart, with an introductory note by Sir Joseph Hooker*. ed. by his sister-in-law Mrs. Henry Lyell (London: J. Murray, 1906), p. 199. This volume contains a similar note Bunbury made later in the year when he met Charles Darwin, and reported that Darwin told him that he was "to some extent a believer in the transmutation of species, though not … exactly to the doctrine either of Lamarck or of the *Vestiges*," p. 213. Also see *Memorials of Sir C.J.F. Bunbury, bart. Naturalist, 1809–1886*, Volume 2, Middle Life (Cambridge: Cambridge University Press, 2011). Bunbury was married to Charles Lyell's sister-in-law.

[19] Letter of January 16, 1840, to Alexander Macleay, written from his residence, 17 Dean Street, Soho. There are several of the letters from Brown relating to his distant cousins, one of whom, John Lyall, wished to travel to Australia and Brown wrote on his behalf to ask Macleay's help and advice. The letters are preserved in the Linnean Society Archives and were transcribed by John Sellick, a fellow of the Linnean Society in 2009.

acquire this magnificent collection but they refused to approve this purchase. The private collector and amateur botanist, Henry Barron Fielding (1805–1851), a fellow of the Linnean Society, purchased Prescott's herbarium, and eventually Oxford University acquired it along with Prescott's books and other material.

Another valuable herbarium, that of Aylmer Bourke Lambert (1761–1842), a fellow of the Linnean Society, became available but the Museum did not purchase it because they were not able or were not willing to provide the necessary resources to pay for it. The collection was auctioned off in 1842, which resulted in the dispersal of important material. The Museum did purchase other collections, e.g., the American plants of Hipólito Ruiz Lopez and José Antonio Pavón Jiménez, the Russian plants of Peter Simon Pallas, Georg Forster's herbarium, Buchanan Hamilton's herbarium from Nepal, and Joseph Martin's collection from French Guiana. This was added to Banks's extensive general herbarium, and the British Museum's herbarium represented the only public herbarium in the London area, and it was under Brown's control. Its access was restricted and until 1853, the British Museum maintained a monopoly over such material. Leading botanists had to develop their own herbaria in conducting their research.

There was no herbarium at the Royal Gardens at Kew when William Jackson Hooker (1785–1865) took over administration of the Gardens in 1841. Hooker brought his private herbarium and library with him when he arrived from Glasgow to take charge of the Royal Botanic Gardens and offered this material to Kew but it could not be accommodated there at the time. Then, Banks's herbarium and library at the British Museum satisfied all Kew's needs. When Hooker was appointed in 1841, Brown, as Keeper of the British Museum Herbarium, made a claim for the important herbarium of plant collectors, Allan Cunningham and James Bowie, a vast collection of plants from South Africa, Brazil, and Australia. Because of the conditions at Kew, this entire collection had been stored in a garden shed at Kew for 16 years and Brown thought that it would be better accommodated at the British Museum.

Hooker realized that he had to improve conditions at the Royal Gardens even before Brown's passing in order to house the herbaria he wished to acquire. Hooker continued his offer to make his own herbarium and library available to the Gardens although he was not entirely happy with this arrangement, i.e., having to make his own private collection available for public use at his own expense. Eventually, in 1847, his collection was housed in a cottage at Kew.

The elder Hooker received the herbarium and library of William Arnold Bromfield (1801–1853), a medical doctor, who lived in the Isle of Wight and traveled extensively. The library consisted of 600 works but the herbarium could not be unpacked because there was no room to accommodate it, a source of some embarrassment. In 1854, George Bentham (1800–1884) presented his vast collection of 103,000 separate specimens, comprising over 50,000 species to Kew. Kew began to revive and grow as a research facility while conditions at the British Museum's

Botany Department became rather static although Brown continued his efforts to acquire new collections of flora.[20]

Brown now was spending the majority of his time working at the British Museum. In addition to botanical matters, he busied himself further by expanding the collection of the "portraits of scientific men" in the British Museum. He wrote a letter to von Humboldt, which was delivered by the London-based photographer, George Henry Polyblank (b.1828, date of death unknown), in Berlin, with the purpose of having von Humboldt sit for Polyblank so he could take a photographic portrait to be added to the collection at the Museum. Brown wrote on October 20, 1857, that "this note will be delivered to you by Mr. Polyblank, a distinguished photographer, who has for some time been engaged in taking Portraits of Scientific Men, mostly Members of the Linnean Club ... is extremely desirous of prefixing yours to that extensive series." He added, "I am happy to hear from several quarters recent and very favourable accounts of your Health, which I should be most happy to have confirmed by yourself. As to myself, now far advanced in my 84th Year, I am still on my Legs, in tolerable health and doing my duty in the British Museum, not ... without infirmities and beginning to lose my Memory of recent events, but am still in comfortable possession of sight & Hearing."[21]

However, age was taking its toll. He was in considerable pain and was suffering from a loss of appetite. Less than a year later, he developed bronchitis and his private physician, Francis Boott, prescribed morphine but Brown did not want to be kept alive in a drugged state. Brown passed away on June 10, 1858, at his home in Soho Square and was buried in Kensal Green cemetery in London. von Humboldt died a year later. In his June 11th letter to Charles Bunbury, Charles Lyell described Brown's last days: "Yesterday morning [June 10th] Robert Brown breathed his last. They told him they might keep him alive till Christmas possibly, Brodie, Bright, and Boott, by opium and stimulants, but he preferred not to live with a mind impaired,

---

[20] William Stearn faults Brown's administration of his duties at the British Museum in the latter part of his life. He cites the deterioration at the British Museum and was more supportive of William Hooker's aggressive efforts after Brown's death. Stearn, "Robert Brown," in *Dictionary of Scientific Biographies*, volume 2, pages 516–522, (1970), and *The Natural History of South Kensington: A History of the Museum, 1753–1980* (London: The Natural History Museum, 1998). A sample of letters preserved in the Linnean Society Archives, to William Macleay, were written on January 20, 1840, where he indicates that he is as "desirous as ever of adding to my New Holland Collection," runs counter to this claim. Another is a letter to Nathaniel Bagshaw Ward (1791–1868), an English physician who was an amateur botanist, and supported efforts to transport collections of plants to museums and herbaria. Brown wrote to Ward on September 6, 1841, in connection with "Price's collection from Swan River" [Australia] indicating that after examining the collection he found "very few novelties" so he was "not inclined to be the purchaser on my own account." But he instructed Bennett "to take charge of it at the Museum & look it over carefully so as to judge of its extent state & the expediency of recommending it to be purchased for the Museum for during my absence this will rest with him."

[21] There is an excellent copy of the entire letter from Brown to von Humboldt (dated October 20/57 and written from his residence on 17 Dean Street, Soho) in John Ardagh's "Portraits and Memorials of Robert Brown of the British Museum," *Natural History Magazine* (1928) 1: 158–162, pp. 161–162.

and so cheerfully and tranquilly, and in full possession of his intellect, gave way to the break up of nature. Every one who has been with him in his last days agrees with me in admiring the resignation with which he met his end, and the friendly way he talked and took leave of us all. Boott has been constantly with him, and looks much worn. Bennett sat up with him many nights."[22]

After Brown's death, his assistant and prodigy, John J. Bennett, became Keeper at the British Museum. Kew became more aggressive in their efforts to acquire the British Museum's collection. Only several days after Brown's death, the Principal Librarian of the British Museum, Antonio Panizzi, indicated in his June 14, 1858, letter to the Lord Commissioners of the British Treasury that there was some pressure to remove the Botanical Collection from the Museum. As a result, a subcommittee was formed to take testimony on the question of whether it was desirable to remove the Collection from the Museum. Ten years previously, a Parliamentary Committee had recommended that upon Brown's death, the Botanical Department at the Museum should be abolished. However, Bennett felt that this action was precipitous because he felt that it ran contrary to Brown's wishes and Brown furthermore left everything in his personal collection to Bennett except his fossils that were given to the British Museum. Certainly, Bennett knew that Brown would not be happy with the possibility of his collection going to Kew and having the Botanical Department at the Museum abolished.

On June 16, 1858, William Hooker testified before a newly formed Committee accompanied by his son, Joseph, as well as John Lindley—Lindley had taken the Professorship at the University of London that Brown had turned down years earlier and continued to harbor antipathy toward Brown even after his death—who was somewhat critical of Brown and his stewardship of the Museums' collections. They urged that the Museum's botanical holdings should be moved to Kew. However, Bennett and Richard Owen testified in defense of maintaining the Museum's collection as it had been previously. The elder Hooker's position was that moving the collection from the Museum was the correct thing to do for its safety (and in spite of duplication because Kew specimens had grown impressively since 1853); it was the logical choice to house the Museum's specimens. Hooker was ultimately unsuccessful in acquiring the Museum's collection, and Bennett kept a firm grip on this material; the Museum's holdings grew, and the Botany Department was not abolished.[23]

Brown considered his own herbarium his private property. When Bennett died in 1876, Brown's private herbarium eventually was incorporated into the collection of what was by now known as the Natural History Museum in South Kensington. Brown's fossil collection and Ferdinand Bauer's drawings had already been acquired

---

[22] Charles Lyell letter of June 11, 1858, to Charles Bunbury, *Life, Letters and Journals of Sir Charles Lyell* 2 vols. ed., by his sister-in-law Mrs. Lyell (London: John Murray, 1881).

[23] See William T. Stearn's *The Natural History Museum of South Kensington: A History of the Museum, 1753–1980* (London: The Natural History Museum, 1998) for a detailed account of this conflict.

Epilogue: The Greatest of Banksian Botanist-Librarians

This book focuses on two Scottish botanists who dedicated much of their lives to widening natural science and exploration, Robert Brown and Mungo Park. They shared similar backgrounds but their careers took quite different turns. They were part of the group of scientists-explorers who contributed to the development of natural history, from the late seventeenth century to the nineteenth century, before Charles Darwin's evolutionary theory was first published in 1858 and 1859. Many of the scientific expeditions, from the late eighteenth century on, emanated from the Royal Society, with Sir Joseph Banks serving as its long-standing President and providing the stimulus for such a monumental string of achievements. There is no question concerning the importance of Robert Brown and why he is the major figure in this work.

But why is Mungo Park included alongside Brown? Park's work in natural history pales in comparison to that of Brown as well as many others whose natural history specimens, illustrations, and observations were sent to Banks. Park deserves more than a passing glance of recognition however, because of his great promise and potential that initially was recognized by Banks. Park indirectly helped Brown's career. If Banks had his way, Park would have gone to Australia with Flinders and might have enjoyed a long and distinguished career as a naturalist. Like Brown, he had an excellent background in natural history, receiving training similar to that of Brown. However, Park's personality was different from that of Brown. His ambitions were focused more on exploration and adventure. His work in natural history during his explorations in Africa took on a decidedly secondary role.

Park had a medical practice in rural Scotland as well as a growing family but his restless spirit and desire to complete his explorations along the Niger River made him leave on such a hazardous journey. Because of the difficulties he endured during his previous trip to Africa, the British Government sent a military contingent to protect him but it was not sufficient. They were fighting "the last battle," preparing for exigencies that occurred previously but Park needed even more manpower. The

J. Schwartz, *Robert Brown and Mungo Park*, Memoirs of The New York Botanical Garden 122, https://doi.org/10.1007/978-3-030-74859-3

limited presence of soldiers merely incited further hostilities ultimately, and the ranks of the armed men were weakened by disease. It was a terrible waste of a talented naturalist and explorer, and others, such as his brother-in-law and close friend, Alexander Anderson, and George Scott, the draftsman for the expedition, also lost their lives. The British Government (Admiralty) seemed to be quite cavalier about these matters, sacrificing people unnecessarily or so it appears.

Compared with Baudin's well-funded expedition, carried out at the same time as Matthew Flinders's journey to Australia, it was remarkable that so much was accomplished. Dedicated naturalists, draftsmen, and navy people like Flinders who spent years in captivity—needless sacrifices—were responsible for making positive contributions. There were other terrible losses; an example was the work and death of Peter Good, a superb collector, who passed away in 1803 around the midpoint of the exploration of Australia. He kept a well-organized diary helpful to Brown as well as historians examining this period. The casual observer might be disposed to claim that the British Government at that time was not particularly interested in scientific investigation but in furthering British colonialism and commercial interests under the guise of scientific exploration. The French, under Napoleon, were doing the same thing but seemed more supportive of the scientific aspects of the effort, with fully equipped seaworthy ships supplied with a sufficient amount of men and material. However, the British were busily engaged in fighting the Napoleonic Wars, and believed that scientific exploration was of secondary importance compared to the survival of their country.

Robert Brown, through his dogged determination, scientific skill, and abilities as an investigator, was responsible for making the circumnavigation of Australia a success, both serving his country and science. It was a critical event in Brown's life and career. It helped him develop as a scientist of the first rank, and mature as a cool-headed individual. Brown forged ahead in many areas of natural history and overlooked the petty annoyances and the backbiting that was directed toward many, including him. This was in contrast to George Caley, a talented collector and naturalist, but one whose letters, particularly to Banks, were filled with complaints and sometimes belittling remarks about his colleagues, including comments about Brown's abilities.

Brown went on to make many important discoveries in the biological sciences that unfortunately have gone on unheralded for too long. When he first arrived in Sydney Cove on May 9, 1802, he did not know that he would spend much of the next three years in New South Wales collecting and organizing the material, and then attempting to bring it back to England. This was in addition to the vast material he collected during the circumnavigation, e.g., from the shores and coves of many areas of Australia. His work in Australia had a profound impact on the natural history and ecology of that largely unknown region.

Brown's influence and eventually his interests went much further than the plants, animals, and geology of Australia. His influence reached into botanical systematics, and his microscopic discoveries made in plant fertilization and cytology were extremely significant. His observation of the movement of suspended particles, named "Brownian movement" in his honor, put him on the threshold of the nature of matter, a subject that was much examined by physical scientists in the late nineteenth century as well as much of the twentieth century.

Brown reintroduced to the English-speaking world a more natural system of classification, borrowing from French systematists as well as other European tax-onomists. He first developed his system in his work on the Proteaceae (a group that includes *Macadamia* trees and *Banksia*, also known as Australian honeysuckle)—a family of flowering plants indigenous to the Southern Hemisphere, particularly Australia and southern Africa. He focused on the Australian species in this family, his first published work that provided the results of his Australian journey. He drew on his knowledge of nearly 200 species of Proteaceae. It proved to be a masterful work, over 200 pages long. Brown classified hundreds of species, placing them in the appropriate genera, 38 in all. He published additional work on other Australian plant species, and wrote an important study on the Asclepiadaceae-Apocynaceae (milkweeds) and Compositae (a group including asters, daisies, sunflowers, chry-santhemum, endive, and chicory).

Brown included work on such families as Sterculiaceae (cocoa and cola nut) and Gesneriaceae (African violets), and described dozens of new families of angio-sperms. He also published work on mosses and ferns, harkening back to the work he did as a youth in the rural lowlands of his native Scotland. He also observed the important distinction between gymnosperms and angiosperms, groups that were previously classified together under the rubric of flowering plants. Through the work of many other botanists and collectors, he studied and classified the specimens that were assembled and sent first to Banks and then directly to him after Banks's death. An example of this was when Brown received the biggest flower in the world, *Rafflesia arnoldii* (corpse lily), a parasitic plant from Sumatra. With diligent study of this unusual plant, he advanced knowledge about this parasitic plant that had no leaves and roots in addition to learning about other parasitic angiosperms.

Up to the time Brown arrived in Australia in 1801, no species from the Rhamnaceae (buckthorns) had been described from there. Several widespread tropi-cal species were already known from elsewhere but Brown collected 31 indigenous species of Rhamnaceae from Australia, primarily from the eastern and southern regions of Australia. He collected additional species from the buckthorn family along the coastal regions, and explored the inland regions in the Sydney vicinity. He found one species of buckthorn from Tasmania and collected representative samples of buckthorns, describing examples of all four of the biogeographical groups in Australia.

In addition to Brown's work on mosses and ferns which reflected his earlier interests begun in Scotland and later reported on in published papers—very much like Park initially—Brown also made other significant discoveries in families of angiosperms, especially Orchidaceae. As already cited, he was responsible for the recognition and establishment of many new families of angiosperms. He examined the pollination and fertilization in flowering plants, particularly in Orchidaceae, uti-lizing his considerable skills with the microscope. He received excellent assistance from Ferdinand Bauer, the draftsman who was so important during his travels and explorations in Australia, and Ferdinand's brother, Franz, later on.

Brown's pollination and fertilization studies were immensely helpful to Darwin. Most importantly, the specimens, drawings, and observations he brought back had a profound impact on naturalists who did not have the opportunity of undertaking voyages of exploration. They opened up new possibilities, demonstrating that the static world they were used to was subject to change.

Toward the middle of the nineteenth century, evolutionary ideas began to gain respectability, and naturalists were forced to consider new ideas so that by the 1840, in the words of Robert Chambers's biographer, they were "in the air."[1] Brown may have allowed for some form of gradual evolutionary change although he did not publish his ideas in this area. His adoption of a natural system of classification served as a framework for accepting evolutionary ideas. The manner with which he rejected the much-revered Linnean (artificial) system demonstrates that he was not timid when it came to these matters. It was not a popular thing to do in Britain but he not only rejected artificial classification, but also proposed a more natural system, taken from the ideas of Jussieu and other European naturalists, and he modified the natural system with the information he personally had obtained, together with the flora that became available to him from distant regions of the globe.

The plants Brown and others collected from Africa were noteworthy—particularly from the Congo basin, Madeira, Western Africa, and Ethiopia—because he observed a similarity between the plants from Africa and those he collected from Australia. Brown's work in plant geography had a significant impact on the work of Humboldt and Lyell. Darwin incorporated this information together with the observations he made during his own travels. Brown also received plants from the Indian subcontinent, North America, and Asia, thereby benefiting from the work of many other talented naturalists.

Brown made discoveries on the nature of the cell, identifying the cell nucleus, thereby adding to the nineteenth-century work that led to the formulation of the cell theory, one of the two important theoretical breakthroughs in the biological sciences in the century. This was made possible by his acute powers of observation as well as his considerable skills in utilizing the microscope.

Brown ranks as a significant investigative naturalist along with Banks who preceded him, and Humboldt, Darwin, Huxley, and Hooker and others who followed him. Johann Reinhold Forster, naturalist on the *Resolution* during Captain Cook's second voyage around the world, was critical of his fellow naturalists including himself. He felt that too often they dealt with miniscule details, confining their study to islands and coasts, and directed their efforts to collecting and compiling lists. Humboldt offered a similar but milder critique, indicating that he thought that they never ventured far from their ships and their researches were confined to "islands and coasts," rather than going farther inland.[2] Brown and Caley spent a good deal of

---

[1] See Milton Millhauser, *Just Before Darwin: Robert Chambers and the Vestiges* (Middletown Conn: Wesleyan University Press, 1959).

[2] Glyn Williams discusses how Johann Reinhold Forster thought he and his fellow naturalists conducted natural history, i.e., "too microscopically, spending their time obscurely rummaging around, counting hairs, feathers and fins." Brown and other naturalist-explorers ventured in often difficult

their time collecting and cataloguing, and they explored the interior regions as well, even if they did not propose revolutionary theories like Darwin's, i.e., tracking "the great and constant laws of nature." Investigators like Brown established important groundwork that Darwin and others utilized.

Robert Brown was a modest individual, a dedicated scientist whose sheer determination compensated somewhat for his retiring manner or shyness. He obviously made the most of his abilities and helped carve out new areas in the botanical sciences, classification, and microscopic studies, which were utilized by later biologists. His remarkable powers of observation were honed by his continual use of the microscope. Where others merely saw, he observed, to paraphrase a line taken from Sir Arthur Conan Doyle's Sherlock Holmes stories where Holmes indicated to his friend and associate, Dr. John Watson, that Watson had overlooked certain important details although he had seen as much as Holmes.[3]

Brown enjoyed his life, fulfilled by his frequent visits to France and Germany, and the companionship of fellow naturalists there. He valued his friends in Britain such as Thomas Smith, Francis Boott (1792–1863), and the American-born physician and botanist, William George Maton (1774–1835), a physician, perhaps his closest friend who wrote on natural history and was a fellow of the Linnean Society. Brown was a friend of Nathaniel Wallich (1786–1854), a surgeon and botanist (born Nathan ben Wulff, in Denmark), who had a particular interest in the flora of India, and of course John Joseph Bennett, his successor at the Natural History Museum.

Brown never married but was interested in women. There is an indication but no direct evidence that he was engaged to Fanny Macleay as early as 1815, daughter of Alexander Macleay. There is the suggestion that at the time, the Macleay family felt that Brown might not have been suitable, feeling that his prospects and age difference might have worked against such a marriage. Later on, the Macleay family may have felt more favorably disposed to the possibility of Brown as a son-in-law but Macleay was appointed to a diplomatic post in Australia and Fanny Macleay went with her family.

Brown was so circumspect about these matters that his letters reveal little about his private affairs. There has been a suggestion that his "housekeeper" in later years, Louisa Harris, was somewhat more than an employee, and that they were romantically involved. She was left a handsome sum in Brown's will.[4] However, other than the generous gift he bestowed upon her, this is merely conjecture for those looking for information concerning Brown's personal life. Short of more tangible evidence,

---

terrain in making their observations and collecting specimens so it is not certain this critique should be applied to Brown and contemporary naturalists. However, Darwin did take von Humboldt's remarks to heart, regarding it as a challenge which he kept in mind when he explored the interior of South America during his five-year voyage on the *Beagle*. *Naturalists at Sea, From Dampier to Darwin* (New Haven and London: Yale University Press, 2013). p. 6.

[3] In "A Scandal in Bohemia," Holmes admonishes his friend, Watson, "You see, but you do not observe. The distinction is clear," *The Adventures of Sherlock Holmes* in *The Complete Sherlock Holmes* (New York: Doubleday & Company, 1927), p. 179.

[4] See *Jupiter Botanicus*, p. 405.

this speculation remains irrelevant in an examination of Brown's life in science. What is relevant is that Brown was a prime example of someone who had the opportunity to travel to distant lands, and benefited from his education and experience in Edinburgh during the Scottish Enlightenment, allowing him to make a significant and lasting impact in the sciences.

Bibliography

1. Ardagh, J. 1928. Portraits and memorials of Robert Brown of the British Museum. *Natural History Magazine* 1: 158–162.
2. Austin, K.A. 1964. *The Voyage of the Investigator 1801–1803, Commander Matthew Flinders.* Adelaide: R.N. Rigby Limited.
3. Bastin, J. 1973. Dr. Joseph Arnold and the discovery of *Rafflesia arnolda* in West Sumatra in 1818. *Journal of the Society for the Bibliography of Natural History* 6 (5): 305–372.
4. ———. 1990. Sir Stamford Raffles and the study of natural history in Penang, Singapore and Indonesia. *Journal of the Malaysian Branch of the Royal Asiatic Society* 63 (259): 1–15.
5. Beaglehole, J.C., ed. 1962. *The Endeavour Journal of Joseph Banks, 1768–1771.* 2 vols. Sydney: Trustees of the Public Library of New South Wales.
6. Botshon, A. 1981. Banks' Florilegium. *Garden* 5: 14–17. (Publication of the New York Botanical Gardens).
7. Broadie, A. 2001. *The Scottish enlightenment: The historical age of the historical nation.* Edinburgh: Birlinn Ltd.
8. Brockway, L.L. 1979. *Science and colonial expansion: The role of the British Royal Botanic Gardens.* New York: Academic Press.
9. Brown, A.J. 2004. *Ill-starred captains, Flinders and Baudin.* Freemantle: Freemantle Arts Centre Press. (Previously published in 2001 by Chatham, London).
10. Brush, S.G. 1968. A history of random process. I. Brownian movement from Brown to Perrin. *Archive for History of Exact Sciences* 5: 1–36.
11. Bunbury, F.H., and K.M.H. Lyell, eds. 2011. *Memorials of Sir C.J.F. Bunbury, bart., naturalist, 1809–1886.* Vol. 2, middle life. Cambridge: Cambridge University Press.
12. Burbidge, N.T. 1960. The phytogeography of the Australian region. *Australian Journal of Botany* 8: 75–211.
13. Cannon, S.F. 1978. *Science in culture: The early Victorian period.* New York: Science History Publications.
14. Chambers, N., ed. 2007–2012. *The Indian and Pacific correspondence of Sir Joseph Banks.* Vol. 1–8. London: Pickering & Chatto.
15. ———., ed. 2008–2014. *The scientific correspondence of Sir Joseph Banks.* Vol. 1–6. London: Pickering & Chatto.
16. Chambers, R. 1842. *The life and travels of Mungo Park and an account of the progress of African discovery.* Edinburgh: William and Robert Chambers.

17. Coats, A.M. 1969. *The plant hunters: Being a history of the horticultural pioneers, their quest and their discoveries, from the renaissance to the twentieth century*. New York: McGraw-Hill Book Company.

18. Darwin, F., ed. 1958. *Autobiography of Charles Darwin, and selected letters*. New York: Dover Publications.

19. ————., ed. 1892. *Charles Darwin: His life told in an autobiographical chapter, and in a selected series of his published letters*. London: John Murray.

20. Dawson, W.R., ed. 1958. *The Banks letters: A calendar of the correspondence of Sir Joseph Banks, preserved in the British Museum, British Natural History Museum and other collections in Great Britain*. London: British Natural History Museum.

21. Delbourgo, J. 2018. *Collecting the world: The life and curiosity of Hans Sloane*. London: Penguin Books. (Previously published in 2017 by Harvard University Press as Collecting the World: Hans Sloane and the origins of the British Museum).

22. Denham, D., H. Clapperton, W. Oudney, and A.V. Salamé. 1826. *Narrative of travels and discoveries in Northern and Central Africa in the years 1822, 1823, and 1824*. 2 vols. London: John Murray.

23. Desmond, R.G.C. 1975. The Hookers and the development of the Royal Botanic Gardens, Kew. *Biological Journal of the Linnean Society* 7: 173–182.

24. Duffill, M. 1999. *Mungo Park, West African explorer*. Edinburgh: Royal Museum (National Museum of Scotland).

25. Edwards, P., ed. 1999. *The journals of Captain Cook, prepared from the original manuscripts by J.C. Beaglehole for the Hakluyt Society*. London: Penguin Classics.

26. Edwards, P.I., ed. 1981. *Journal of Peter Good: Gardener on Matthew Flinders voyage to Terra Australis 1801–1803*. London: British Natural History Museum.

27. ————. 1976. Robert Brown (1773–1858) and the natural history of Matthew Flinders' voyage in H.M.S. Investigator, 1801–1805. *Journal of the Society for the Bibliography of Natural History* 7: 385–407.

28. Flannery, T., ed. 2000. *A voyage to Terra Australis: Matthew Flinders' great adventure in the circumnavigation of Australia*. Melbourne: The Text Publishing Company. (A Voyage to Terra Australis originally was published in 1814 in two volumes and an atlas in the third. G. & W. Nicol, London).

29. Flinders, M. 1805. Concerning the differences in the magnetic needle, on board the investigator, arising from an alteration in the direction of the ship's head. *Philosophical Transactions of the Royal Society* 95: 186–197.

30. Gilbert, B. 1984. The obscure fame of Carl Linnaeus. *Audubon* 86: 102–114.

31. Gilmour, J. 1944. *British botanists*. London: William Collins.

32. Gwynn, S. 1934. *Mungo Park and the Quest of the Niger*. London: John Lane, The Bodley Head Limited.

33. Hakluyt, R. 1588. *A briefe and true report of the new found Land of Virginia*. London: Principall Navigations.

34. Hatfield, G. 1981. *Robert Brown A.M. F.R.S.E. (1773–1853). Scottish men of science*. Edinburgh: Edinburgh University Press.

35. Hewson, H. 2002. *Brunonia Australis: Robert Brown and his contributions to the Botany of Victoria*. Canberra: The Centre of Plant Diversity Research.

36. Hooker, J.D., ed. 1896. *Journal of the Right Honourable Sir Joseph Banks made during Captain Cook's First Voyage on H.M.S. Endeavour in 1768–71 to Terra del Fuego, Otahite, New Zealand, Australia, the Dutch East Indies*. London and New York: Macmillan & Co. Ltd.

37. Horner, F. 1987. *The French Reconnaissance: Baudin in Australia 1801–1803*. Melbourne: Melbourne University Press.

38. von Humboldt, A. 1997. *Cosmos. A sketch of a physical description of the universe*. 2 vols. Baltimore: Johns Hopkins University Press.

39. Ingleton, G.C. 1986. *Matthew Flinders: Navigator and Chartmaker*. Australia & Hedley, Surrey: Genesis Publications.

40. Jenkinson, J. 1993. *Scottish Medical Societies, 1731–1939, their history and records.* Edinburgh: Edinburgh University Press.
41. Johnston, H. 1912. *Pioneers in West Africa.* London: Blackie & Son, Ltd.
42. Jonsson, F.A. 2013. *Enlightenment's frontier, the Scottish highlands and the origins of environmentalism.* New Haven & London: Yale University Press.
43. Jussieu, A.L., de 1789. *Genera Plantarum: Secundum Ordines Naturales Disposita, Juxta Methodum in Horto Regio Parisiensi Exaratam, Anno 1774. (Genera of Plants Arranged According to Their Natural Orders, Based on the Method Devised in the Royal Garden in Paris in the Year 1774).* Parisiis: Apud viduam Herissant et Theophilum Barrois.
44. Kellermann, J. 2004. Robert Brown's contributions to Rhamnaceae systematics. *Telopea. A Journal of Plant Systematics* 10 (2): 515–524.
45. Knight, J. 1809. *On the cultivation of the plants belonging to the natural order of Proteaceae, with their generic as well as specific characters and places where they grow wild.* London: W. Savage, Printer.
46. Lipkowitz, E.S. 2014. Seized natural-history collections and the redefinition of scientific cosmopolitanism in the era of the French Revolution. *The British Journal of the History of Science* 47 (1): 15–41.
47. Lupton, K. 1979. *Mungo Park, the African Traveler.* Oxford: Oxford University Press.
48. Lyell, K.M.H., ed. 1906. *The Life of Sir Charles J.F. Bunbury, bart., with an introductory note by Sir Joseph Hooker.* London: J. Murray.
49. ———., ed. 1881. *Life, letters and journals of Sir Charles Lyell.* 2 vols. London: J. Murray.
50. Mabberley, D.J. 1985. *Jupiter Botanicus, Robert Brown of the British Museum.* London: British Museum (Natural History).
51. Miller, D.P. 1983. Between hostile camps: Sir Humphry Davy's Presidency of the Royal Society of London, 1820–1827. *The British Journal for the History of Science* 16 (1): 1–47.
52. Millhauser, M. 1959. *Just Before Darwin: Robert Chambers and the Vestiges.* Middletown CT: Wesleyan University Press.
53. Morrell, J. 1974 & 1997. Reflections in the History of Scottish Science. *History of Science* 12: 81–94. (Republished in Science, Culture and Politics in Britain, 1750–1870. Aldershot, Hampshire: Variorum).
54. Musgrave, T., and W. 2000. *An empire of plants: People and plants that changed the world.* London: Cassell & Co.
55. O'Brian, P. 1993. *Joseph Banks: A life.* Boston: David R. Godine.
56. Peck, R.M. 2014. Discovered in Philadelphia: A third set of Thomas Horsfield's nature prints of plants from Java. *Archives of Natural History* 44 (1): 168–170.
57. Peck, T.W., and K.D. Wilkinson. 1950. *William Withering of Birmingham.* Baltimore: The Williams and Wilkins Co.
58. Rice, T. 1999. *Voyages of discovery: Three centuries of natural history exploration.* New York: Clarkson Potter, (for the Natural History Museum of London).
59. Richardson, J. 1846. *Ichthyology of the voyage of H.M.S. Erebus and Terror.* London: Longman, Brown, Green, and Longmans.
60. Rigby, N. 1998. *The politics and pragmatics of seaborne plant transportation, 1769–1805. Science and exploration in the Pacific European voyages to the southern oceans in the eighteenth century.* Woodbridge, Suffolk and Rochester, NY: The Brydell Press in Association with the National Maritime Museum.
61. Robinson, T. 2008. *William Roxburgh: The founding father of Indian Botany.* Edinburgh, Chichester: Phillimore, in Association with the Royal Botanic Garden.
62. Ross, J. 1819. *A voyage of discovery made under the orders of the admiralty, in his Majesty's Ships Isabella and Alexander for the purpose of exploring Baffin's Bay and inquiring into the probability of a north-west passage.* London: John Murray.
63. Rourke, J.P. 1974. Robert Brown at the Cape of Good Hope. *Journal of South African Botany* 40: 47–60.

64. Salt, H. 1814. *A voyage to Abyssinia, and travels into the interior of that country, executed under the orders of the British Government, in the years 1809 and 1810*. London: Printed by F. C. and J. Rivington, by W. Bulmer and Co.

65. Schiebinger, L.L. 2004. *Plants and empire: Colonial bioprospecting in the Atlantic world*. Cambridge, MA and London: Harvard University Press.

66. Schulz, M.C. 1805. *Mungo Park's Reise in Afrika: für die Jugend bearbeitet*. Berlin: Schüppelsche Buchhandlung.

67. Schwartz, J. 1999. Robert Chambers and Thomas Henry Huxley, science correspondents: The popularization and dissemination of nineteenth century natural science. *Journal of the History of Biology* 32: 343–383.

68. Scoresby, W. 1820. *An account of the Arctic Regions, with a description of the northern whale-fishery*. Vol. 2. Edinburgh: A. Constable.

69. Shapin, S. 1974. The audience for science in eighteenth century Edinburgh. *History of Science* 12: 95–121.

70. Smith, C.H. 2018. Alfred Russel Wallace note 8: Wallace's earliest exposures to the writings of Alexander von Humboldt. *Archives of Natural History* 45 (2): 366–369.

71. Smith, E. 1911. *The life of Sir Joseph Banks, President of the Royal Society with some notices of his friends and contemporaries*. London and New York: John Lane Company.

72. Smith, J.E. 1811. An account of a genus of New Holland plants named *Brunonia*. *Transactions of the Linnean Society of London* 10: 365–370.

73. Smith, Lady P.R., ed. 1832. *Memoir and correspondence of the late Sir James Edward Smith*. London: Longman, Rees, Orne Brown, Green and Longman.

74. Stearn, W.T. 1960. Franz and Ferdinand Bauer, masters of botanical illustration. *Endeavour* 19: 27–35.

75. ———. 1976. *The Australian flower paintings of Ferdinand Bauer*. London: The Baslick Press.

76. ———. 1998. *The Natural History Museum of South Kensington: A history of the museum, 1753–1980*. London: The Natural History Museum.

77. Thomson, J. 1890. *Mungo Park and the Niger*. London: G. Philip & Sons. (1970 Argosy Reprint).

78. Tuckey, J.H., and C. Smith. 1818. *Narrative of an expedition to explore the river Zaire usually called the Congo in South Africa, in 1816*. London: John Murray.

79. Watts, P., J.A. Pomfrett, and D. Mabberley. 1997. *Ferdinand Bauer (1760–1826), an exquisite eye: The Australian Flora & Fauna Drawings 1801–1820 of Ferdinand Bauer*. Glege, NSW: Historic Houses Trust of New South Wales.

80. Webb, J. 1995. *George Caley, nineteenth century naturalist: A biography*. Chipping Norton: S. Beatty & Sons.

81. Welebit, D. 1988. The "Wondrous Transformations" of Maria Sibylla Merian. *Garden* 12: 11–13. (Publication of the New York Botanical Gardens).

82. Williams, G. 2013. *Naturalists at sea, from Dampier to Darwin*. New Haven and London: Yale University Press.

83. Willis, J.H. 1956. Robert Brown's collectings in Victoria. *Muelleria* 1: 51–54.

84. Wood, Paul. 2003. *Sciences in the Scottish Enlightenment. The Cambridge companion to the Scottish Enlightenment*. Cambridge: Cambridge University Press.

85. Wulf, A. 2009. *The brother gardeners: Botany, empire and the birth of an obsession*. New York: Alfred A. Knopf.

86. ———. 2015. *The invention of nature: Alexander von Humboldt's New World*. New York: Alfred A. Knopf.

Books and Papers by Brown and Park

87. Bennett, J.J., ed. 1866–68. *The miscellaneous botanical works of Robert Brown.* London: Ray Society for R. Hardwicke.

88. Brown, R. 1828. A brief account of microscopical observations made in the months of June, July and August, 1827, on the particles contained in the pollen of plants; and on the general existence of active molecules in organic and inorganic bodies. *Edinburgh New Philosophical Journal* 5: 358–371. and Annales des Sciences Naturelle (Paris) 14: 341–362. (Privately printed copy is in the LuEsther T. Mertz Library of the New York Botanical Gardens).

89. ———. 1821. An account of a new genus of plants named *Rafflesia. Transactions of the Linnean Society* 13: 201–234.

90. ———. 1824. *A supplement to the appendix of Captain Parry's Voyage for the discovery of the north-west passage in the years 1819–20.* London: John Murray.

91. ———. 2001. *Diary of Robert Brown, British Museum (Natural History, Botany Library).* Published as Nature's Investigator; the Diary of Robert Brown in Australia, 1801–1805, compiled and edited by T.G. Vallence, D.T. Moore & E.W. Groves. Australian Biological Research Study, Canberra.

92. ———. 1814. *General remarks, geographical and systematical, on the Botany of Terra Australis. M. Flinders, a voyage to Terra Australis; undertaken for the purpose of completing the discovery of that vast country, and prosecuted in the years 1801, 1802 and 1803, in His Majesty's Ship the Investigator.* 2 vols. London: Atlas G. & W. Nicol. (Reprinted in the Miscellaneous Botanical Works of Robert Brown. J. J. Bennett (ed.). 1866–68. Published for Ray Society for R. Hardwicke, London).

93. ———. 1817. *Observations on the natural family of plants called Compositae.* London: Richard and Arthur Taylor. (Extracted from the Transactions of the Linnean Society of London 12: 76–142).

94. ———. 1833. Observations on the organs and mode of fecundation in Orchideae and Asclepiadeae. *Transactions of the Linnean Society of London* 16: 685–745.

95. ———. 1811. On the proteaceae of Jussieu. *Transactions of the Linnean Society of London* 10: 15–226.

96. ———. 1810. *Prodromus florae novae Hollandiae et Insulae Van Diemen (Prodromus of the flora of New Holland and Van Diemen's Land).* London: Richard Taylor & Son.

97. ———. 1811. Some observations of the parts of fructification in mosses; with characters and descriptions of two new genera of that order. *Transactions of the Linnean Society of London.* 10: 312–324.

98. Park, M. 1797. Descriptions of eight new fishes from Sumatra. *Transactions of the Linnean Society of London* 3: 33–38.

99. ———. 1840. *The life and travels of Mungo Park, with the account of his death from the Journal of Isaaco, the substance of the later discoveries relative to his lamented fate, and the termination of the Niger.* New York: Harper and Brothers.

100. ———. 1799. *Travels in the interior districts of Africa.* London: W. Bulmer and Co.

101. ———. 2002. *Travels in the interior of Africa.* Ware, Hertfordshire: Wordsworth Editions Limited. (Reprint of Travels in the interior districts of Africa, W. Bulmer and Co., London, with additional material).

102. Whishaw, J., ed. 2012. *The journal of a mission to the interior of Africa in the year 1805, together with other documents, official and private, relating to the same mission, to which is prefixed an account of the life of Mr. Park.* Adelaide: The University of Adelaide Press.

Index

Printed in the United States
by Baker & Taylor Publisher Services